Wander

By Pete and Luisa.

A story of personal growth, empathy, romance, and the harsh realities of a year long horseback adventure on the Asian steppe.

Table of contents.

Introduction. Page 5

Chapter 1. Misfits. Page 7

Chapter 2. A leap of faith. Page 22

Chapter 3. Sunny days. Page 38

Chapter 4. It takes 100 falls to make a cowboy. Page 57

Chapter 5. Never alone. Page 96

Chapter 6. Becoming a Horseman. Page 125

Chapter 7. Into the wild. Page 150

Chapter 8. Saddles, girths and tack. Page 167

Chapter 9. Balance and total weight limitations. Page 181

Chapter 10. Survival, field craft and navigation. Page 193

Chapter 11. Reality check. Page 213

Chapter 12. Letting go. Page 232

Acknowledgments. Page 244

Introduction.

"Once more into the frey.
Into the last good fight I'll ever know.
Live and die on this day.
Live and die on this day..."
The Grey

The twists and turns that led a young woman and her faithful dog to ride across Asia with a complete stranger are complex, and not far off the script of a hastily slapped together sitcom. Pete and Luisa could not be more opposite; a wild and obnoxious Australian hunter with a dark past, and a bubbly, free spirited former vegan and trained chef from Bavaria.

United by their love of horses and dogs, they set off unprepared, under-equipped and way out of their depth just days after a chance meeting in a Mongolian park. It's a journey full of personal failures and triumphs; romance, and dreams shattered, lost, re-claimed and rebuilt only to fall apart once more. But if I am to start anywhere, then perhaps I start by saying we wrote this book for people just like us. So, if you do contemplate taking any sort of personal journey of self-discovery, you do it with a chance to avoid many of our failures.

We hope these stories both inspire and educate others that dream of doing something they can look back on and smile about in years to come... smile knowing they did something reckless, beautiful and inspiring; and that even in failure, there can be success. This is a book written by a dreamer and a drunk, who's ambitions exceeded their abilities to such a degree... magic happened daily amid the chaos.

Telling the story of our adventure chronologically seems wrong to us; instead, we're excited to share the lessons that taught us the most, and offer a glimpse at our feelings and emotions, as well as advice on how to take your own crazy trip. If you are brave, empathetic, physically fit and a little crazy... then go make your own mistakes like we did.

We aren't special, we're just normal people that decided life was too short. At first this was a journey we took as individuals, then as a team, and although we at first hadn't planned to write about it, more and more often we have been asked to share our adventures and give a few pointers.

Right from the start we recorded our trip each few weeks on YouTube, not to get famous, sponsored, or for money; but because we wanted to be able to share and re-live our journey with our friends and family, both at the time, and in the years to come. Our journey is far from over, and as I type the final chapters, we are mere days from crossing into Turkey to continue our ride towards Europe.

If you would like to follow our journey from here, or look back over our adventure thus far, the Acknowledgements contains links to where to find our videos and stories, as well as links to the amazing people, places and equipment that have made the trip such as success.

Chapter 1.
Misfits.

"I haven't failed, I just found ten thousand ways that don't work."
Thomas Edison.

It's not easy to start telling this tale, and it seems it's best to start in a metaphorical sort of way for things to make sense. You see; this isn't a book about horses, riding, or even us. It's a book about living and loving life; all things in it, both good and bad. But this first chapter is about two people; two people born worlds apart who met like dry soil sewn with the seeds of adventure, and warm spring rain. What sprung into life was a bond forged in such bright sunlight, that the dark rainy days that often followed only served to strengthen our growth when again the sun shone.

Pete.

My names Pete, and planning a long ride started for me some years ago. It must have been some point in 2015 when the idea of just riding off into the sunset started to become more appealing than continuing a normal life in New Zealand. I'd just left a series of very traumatic experiences in the jungles of the South Pacific behind me, and was attempting to piece together some sort of a future at home with my partner. So, what kind of work do you do after a life like mine?

Let's be honest here, there's not really anything I'm suited to in the real world, and I had pretty bad anxiety at the thought of being forced to fit in. My world was just too different and my return to so called normality was anything but normal. My relationship with an awkward but lovely English girl was anything but normal as well.
Our relationship started in a hut on my farm where I guided hunting trips, deep in the hills of the Wanganui region of New Zealand's north island in mid 2012.

Things were hard on us both, particularly as I was mid-way through an awful divorce, and her romantic ideas regarding the adventure we would have together were just as misguided as mine. But she had a magic about her, and awkward beauty and strangeness I loved despite us being absolutely terrible for one another. I was a wild extrovert tormented by my experiences overseas, she was a quiet introvert that liked nothing more than snuggling up on the couch watching cartoons.

So, after everywhere I've been and everything I've done, seen, lived and fought through; I'm reduced to this... spinning a stop/go sign at roadworks working for a temping agency. I wasn't the only guy like me there spinning a stop/go sign. On one of my first days on the job I met a veteran of the war in Afghanistan. He hadn't been back very long, and had been out of military service a matter of weeks when we met on a rainy day in the suburbs. We were working on a steep side street, with well-kept lawns and neat rows of houses clearly loved and maintained beautifully by proud occupants; families that loved one another, and wanted for little.

A mini digger worked between us, digging up a small section of foot path where a small water leak had been causing the residents a mild inconvenience. This place was different to the lands we dreamed of each night, and the people so different to the ones we had fought to protect and offer even just a hint of the freedom we enjoyed here and so many took for granted.

He drove a 5 million dollar, 22 ton armored fighting vehicle a year earlier, and now he spun a sign back and forwards to let traffic up and down a side street. A few months earlier I was negotiating gold deals "often at gunpoint' while sipping a warm beer among the crumbling remains of the Panguna mine in central Bougainville... and now here we where. Anxious, afraid of what the future holds for us and unable to communicate with anyone, let alone those closest too us. So this was it... minimum wage, spinning a sign from stop to go for 8 hours a day 5 days a week giving us endless hours to contemplate how far we had fallen from a time when we were something - something that mattered - made a difference and could affect real change in the world.

A mum stood near us with a young boy in a stroller who was excitedly watching the digger work... I remember being so envious of that young boy, and a world he still viewed with wonder. I didn't like being home and nothing felt right here.

Things smelt wrong, felt wrong, and I just can't express in words how strange it is to be home, when home doesn't feel like home anymore. When you stand in the shower until the hot water runs out, just standing still, letting water as hot as you can bare just fall on your neck. Feeling it press softly against the giant knot of tension you carry every day, thinking of all those months without a hot shower. Thinking of all the nights unable to sleep lying in the dirt and rotting leaf litter of the jungle floor as you dry yourself with a soft scented towel, in a clean bathroom that smells of citrus soap. I would stand there, watching the fading steam move about the room just as it would in the dim light filtering through the canopy of the forests that still haunted my every waking moment.

Laying down in bed, smelling the fabric softener on the sheets and rolling over to press against my partner, running my hands over her curves and smelling her hair. It felt like a dream I would have to wake up from at any second; and there was more than one occasion where I woke up terrified with no idea where I was.

I felt isolated working as a temp, and my inability to relax was very difficult for both my partner and I to come to grips with. Every day my behavior pushed me further from others, and I didn't care. I didn't want to be anything like them, these youth, these millennial types all in a state of prolonged adolescence. Students too scared to enter the work force and grow up, kids that couldn't hold down a job through drug, alcohol issues or just laziness.

They all felt they were owed something, and at times it was like being back in the Solomon Islands, Bougainville, PNG and Indonesia where every day held a new risk and challenges as I hustled making a living in the gold trade. As the days passed, I found temping more and more difficult, and as the weeks faded into months, I hit breaking points; each more dramatic than the last. I needed to find a way out, and all those crazy times in dangerous places kept calling me back each night in my dreams.

I made a reasonable wage through, mainly from working ridiculous hours, and it was enough to allow my partner to get back into riding a little; something I took very great pride in as it was always her childhood dream to own a horse. It didn't take her long to form new friendships in the horse community, and I would often come along to watch.

Truth is I loved horses, I always have, and sharing in the fulfillment of her dream gave me a little of that purpose I so desperately sought. I was soon a big part of it, working on the dilapidated riding school fences each weekend in exchange for free riding lessons for my partner.

I loved every minute of it, just being around horses with my partner and our dog each weekend was enough to settle me, and start showing me there could be options at home where I could find peace. I started with counseling more regularly too, seeing a psychotherapist and touching on the issues relating to my PTSD, but most importantly, my mother. I'd "ruminate" in his words, continually chewing over in my mind things I needed to let go of and get on with. Drinking each evening was the only thing that got me to sleep most nights, and strangely he felt my drinking actually didn't do me any harm, clearly stating it's not a problem unless you cause problems.

With the success of fencing at the riding school, I soon found myself fencing full time for a fencing contracting company, in no time, I was clashing with my co-workers. I wanted to move both the business and my role in it forward, to see some real money coming in and the business grow. Those I worked alongside just wanted pay at the end of the week, and to crack back on with work Monday doing the absolute bare minimum. My partner pushed me to stay, wanting to buy a house and urged me to keep that regular income needed for a mortgage; but I couldn't do it. The boss was cool, but my supervisor was a "muppet" and watching the constant waste of hours, materials and effort just about brought me to tears.

After watching what I calculated to be $500 wasted in a morning, I snapped, said exactly how I felt about things, quit, and began fencing myself, on my own term for my own clients.

It was a smart move, and in no time I had built a very successful little business for myself and partner able to fund all of her horse dreams. Although the new business would have its challenges, it offered me more of a chance to feel like I did in those paddocks at the riding school, watching my partner potter around with the horses while I strained and repaired fences. I was walking my own path once more; and for my partner, although very reluctant and somewhat annoyed when I quit legitimate employment, the money I was able to bring in soon had her more than happy with me.

But of course I grew restless, and the offer of a quick security job or two once more had me back where things made sense. Between guiding hunting clients, the security jobs overseas and my fencing, I was finding some sort of balance in life. Boring nine to five, house, dog and horses on the weekend, and still fulfilling my need for adventure in crazy foreign lands... but it couldn't last.

I didn't have a plan then, but one was forming slowly. Pieces were falling into place, and I was realizing more and more the company I kept would ultimately make or break things for me.

Throwing myself again and again into lost causes in support of my partner, and my ex-wife and mother were doing nothing but tear me apart from the inside out. Worse still, the person I was as a result of these interactions is one I didn't like. Anxious, tired, stressed and constantly in physical pain from over working and under sleeping, even with my nightly half a dozen beers or bottle of wine.

I tried changing the way I worked, yet more and more I was realizing that I was doomed to failure trying to be something I'm not, for people whose investment in me did not match my investment in them. It was time to start saying "Ima be me" and mean it. I knew it would mean leaving my partner, but we couldn't keep on destroying each other with our incompatibility.

I need to keep walking the hills, keep patting puppies and ponies and allow those invisible friends that have stalked silently beside me for so long come out from the shadows and walk with me in the light.

My time in the wild, my time alone with my thoughts balances me and helps me find a way forward in hard times. The friends I've made, the things I've seen and even those people I alienate myself from only serve to strengthen me in my search for happiness, peace and a sense of being in a world that stresses and confuses us all at times. A bond with the wild, communicating effortlessly with those who don't share a common tongue, sharing your warmth with a dog by a fire on a cold night and working with a wild horse you bond with are feelings that come to hunters and those that inhabit wild places.

Feelings are something men don't talk about often. But I feel more whole, more at peace and happier when deprived of comfort, alone in the wild with others who feel what I feel than I could ever feel sipping a latte in a hipster filled cafe.

It was these feelings that drew me to Outer Mongolia in search of peace. An endless floral pasture with a scent all of its own; dotted with wandering mobs of sheep, goats and yaks. Nomadic camps scattered loosely along the horizon, while mobs of horses graze quietly seemingly everywhere you look. Ground squirrels scamper between there numerous little holes, and I stand in silence sipping a bowl of breakfast vodka, eating a wolf meat dumpling; watching the sunrise in peace. I was here on unrelated business, but opted to just wander off into the northern Mongolian forests in search of adventure, wolves to hunt and new experiences with new risks and reward, and it was here I met my horse "Digger".

My eyes have just welled up with tears thinking of the time I've shared with him on that first, and subsequent trips. When we first met, Digger was a 4-year-old gelding of around 13.2 hands. He was kind, friendly, powerful but very aggressive, and ran down and killed a wolf that was attempting to kill younger horses he ran with. He "thought" with me if that can possibly make sense? He and I shared a bond than can only really happen when you let go of everything, and it was with him I first coined the phrase...

"I hope for nothing, fear nothing, and I am free."

Words that sum me up perfectly, and when Digger and I were together magic happened. I was alive in ever perfect moment we shared. I feared nothing, galloping with a slung AK47 listening to Tupac "Picture Me Rolling" on my iPhone with crazy Mongolian nomadic horsemen laughing and cheering as we rode blind drunk through the untouched northern wilderness bellow Lake Baikal. I was free. Completely free! Fearless, and nothing could make any moment we shared more perfect than it was.

In the days that followed I charged 3 bears on horse-back at the gallop, caught the mighty Taimen "the world's largest type of trout", foraged food from the forest and drank from crystal clear steams. I sang to my pony and hugged him blind drunk lying in the soft summer grasses under a star filled sky thousands of kilometers from anything you'd call "civilized". This was what I was looking for my whole life. I was free from anxiety, fear and without a care in the world. I ate marmot, the last animal known to carry the black plague. I laughed. I cried, and I lived as man was supposed to.

Sadly, as soon as I was home I struggled with the constant voices of those "in the know" telling me what I should be doing. Everything I was doing was wrong apparently; I ride with my seat, use my inside leg with my inside rein, and having seen what a horse is capable of ask them to do things considered ridiculous by those that do laps around an arena on an off the track thoroughbred atop a $3000 English saddle.

But I'd learned a lot about how to have a relationship with a horse, and although this was all deemed wrong by my partner, I'd come to love Digger and understand horses in a way she never would, regardless of how poorly others thought my technique was. Digger was like a missing piece of my soul that fell in place every time I looked into the soft brown eyes of that beautiful little plain bay pony. But the most important lesson he taught me was this... horses have a better understanding of empathy than most people; and it was time for me to leave my life here behind for good.

I had what I can only describe as an unequivocal adventure that first horseback trip in Mongolia, and the bonds I formed with my brothers will last a lifetime. The bonds men form in combat, in wild places and in adversity while measuring themselves physically and emotionally are unlike any other.

Did it make me a horseman? I'm not sure, but I began to understand that my journey as a horseman would be one I would never walk a single day in without evolving as a rider and a man.

It was there in the land of blue sky where I finally glimpsed a chance at the lasting peace I'd so desperately sought, and it was there that I met the two greatest loves of my life some 3 years later. Their names are Luisa and Jill, and this is where their story begins.

Luisa, and Jill "the worst cattle dog ever"

Luisa is a real character; and as I type, she is by my side as she has been almost every minute of the last year. This is our story; but Luisa is Bavarian, and her spoken English is a cosmopolitan mix of Australian and Kiwi slang, swear words, cutesy loving baby talk with her dog, hysterical incoherent laughter... and German.

Although Luisa and I have no problem communicating, for our story to work I'm going to predominantly do the typing for us both. This is our book, written by us together, and these are our words, but Luisa and I have decided it's best I write primarily from my point of view. Any story written in its entirety by Luisa, will be in *Italics* to hopefully avoid any confusion.

It's only now as Pete and I sit down to write in the shade of a sprawling grape vine, heavy with deep purple Sapevari grapes on the deck of our little Air bnb apartment in Tbilisi, I realize just how long I've been on this path. This crazy winding journey that has me, a 25-year-old girl in love with a life born of bold dreams and blind leaps into uncertainty.

In my younger years I was a bit withdrawn and both Mum and Dad worried about me, as I had some pretty unreasonable fears. I had the smallest room in the house by choice, and feared imaginary monsters lurking in dark corners, or rapists jumping in the window at night. I was shy, and spent my days hiding away, losing myself in books and my imagination. Happy in my own skin and head despite my irrational fears, I was quite different to my brothers and sister, but that's what made us such a great team.

My older brother and I were inseparable; he was so full of adventure, and in some ways what I had in fears, he had the opposite of in courage. As a result, Dad had to rescue us more than once.

One time we drifted out into the open ocean in our little inflatable boat while holidaying in France. The current took us, and we ended up hanging onto a buoy for dear life until rescued by Dad and the coast guard.

Or when we ran through a paddock, screaming and laughing, full of childish bravado as I followed my big brother, unaware the stable we planned to play in towards the middle of the paddock was not really a stable... but a bull's shelter. Moments later, it had us well and truly cornered there, just my brother and I, and an angry bull with nowhere to go until dad saved us again.

My little brother on the other hand, he figured out I just wouldn't say no to his charms. So, he took advantage of me; Sunday morning, 7 o'clock, he'd come into my bedroom, a board game under his arm. His blond hair falling into his eyes, covering his soft face, never willing to leave until I'd played with him. But we couldn't stay young forever, and as my big brother started to grow up, getting interested in girls and so on, sadly, he started to have better things to do than hang out with me.

When he brought home his first girlfriend, my little brother and I deiced we had to inspect the situation from a secure angle first. The smell of Mum's lasagna filled the living room and kitchen as we crawled on our hands and knees to spy on the lovers. As we entered the kitchen, Mum stopped us with an irritated look. I looked around with a serious look on my face at my little brother, knowing our secret mission was a bust. Mum asked if we were pretending to be dogs... so I just started barking, hearing laughter long before I saw the faces of a tall gorgeous blonde girl and my irritated brother. I froze just for a second, then jumped up and ran, only to be called straight back for a lunch I spent red faced a silent.

My sister was more girly than me and she still is to this day, I'd argue the more beautiful one of the two of us as well, I often wished I could be more like her. She was so headstrong, so true to herself and so loud in her opinion's. She was the only one I really fought with, and there was nothing I hated more, as I just wanted the world to be happy and peaceful.

She often reminded me to be a girl, and when forced to watch "girl shows" on TV, I would simply join her and sort of switch off. All my siblings were excellent at gymnastics too, and even though I tried for years, I disliked the group pressure, the upcoming events, and the competitive mindset required which I simply didn't have in me. I started to look for something to make my own, and found it when a friend dragged me along on a "pony holiday camp".

I was 11 years old, and a week without the always embarrassing parents, and a reminder of recently barking like a dog on the kitchen floor seemed like heaven.

Being around a horse for the first time fulfilled me in ways I had only imagined, and although my hand was a little shaky the first time I touched a horse... I soon wanted for nothing else. I learned nothing about horses; absolutely nothing, but I tell you what... it was the beginning of a love that will be with me for the rest of my life. The nights sleeping in the hay, and the days filled with brushing and riding ponies left me with a smile like never before, and a lust for more time with horses with regular riding lessons the inevitable result.

The next few years my friend and I learned how to catch, saddle, and ride horses. Each night as I drifted off to sleep, I'd count the days until the next riding lesson while my imagination ran wild with dreams of horseback adventures and mustering cattle. When I closed my eyes, the houses, streets, and people disappeared; endless space, forests and mountains replaced them and I was galloping a wild black stallion with a big loyal dog by my side. The wind in my hair, freedom, and nothing but the setting of the sun leading the way towards tomorrow and whatever it may bring us.

And when I did have to open my eyes again... I took some of the nights dream's and magic with me, my bicycle turned into a horse, and I would talk to it, knowing it understood me; I would bring him grass, and sometimes even a carrot. Horses became my secret in some way, and my obvious passion in another. My parents tried to support me by coming by the riding lessons at first; but I didn't want that, I really didn't want to talk about it or feel the pressure of being watched.

I vividly remember my parents asking me to sit down with them for a chat; I wasn't sure what I did wrong. I went through the past few weeks in my head... nothing happened, I hadn't fought with my siblings, school was good so what was it? I always felt very anxious at our table conversations, with it being two against one, me on one side, my parents on the other. We had a big wooden table, always with flowers and a big old woolen carpet beneath it. It was a bit scratchy for bare feet, but fitted the room nicely and it was here my life again took one of those turns that steered me towards today's dream like setting under the grapes in the late summer sunshine.

They looked at each other, both starting to smile... they'd agreed it would be best for me to have a horse to ride and take care of in exchange for my helping out in a nearby riding stable. Momentarily speechless, I jumped up hugged them, barely even able to take a breath as I thanked them over and over again. My whole body was fizzing, so unbelievably happy that before they could put any conditions on it, I darted out of the room, not sure if I could wait for my next riding lesson and to tell my riding teacher the good news!

But I didn't start off with a suitable horse; a stubborn mare not even used for riding lessons and I simply wasn't good enough for her. It didn't matter how hard I tried, every week it got more dangerous. Even so I didn't really want to give her up, not after I just had my parents convinced, but luckily life presented me with opportunity once more. My older brother did a school exchange program to Australia, and a handsome young man by the name of Luke came to stay in place of my older brother Mo. At first... this shy Bavarian girl had a bit of a crush on the stranger from Australia now living as part of my family.

His German was good, and my English was basic at best. Soon, that 16-year-old schoolgirl crush was replaced with confidence and trust that rivaled the way I felt about my own brother, and with both my confidence and ability to speak English developing... an opportunity presented itself. Luke had a younger sister my age, and with Luke returning to Australia, I had a chance to exchange with her, while staying safely under the wing of a kind, caring German speaking young man that had come to feel like a brother to me. I jumped at the chance, and had 3 amazing months in Australia that was over all too soon.

I was 16 now going on 17, and with my thoughts still in Australia, I got a chance on my return to Bavaria with a new horse. A 3-year-old gelding; a beautiful paint just broken to saddle, and a calm, gentle character with potential to be a superstar. I had so much fun with him mooching though the forest, brushing him for hours and working in the round pen. I was absolutely crazy about him, and recall bursting into tears as an emotional teen unable to get him to listen to me while doing ground work in the round pen. I was frustrated; and all I wanted to do was throw away the rope and walk away.

Self-doubt flooded in, crushing my dreams of becoming any sort of horseman and I was sure that stunning paint would be better off without me. Maybe I just give up on my childish horse dreams and move on? I didn't know what to do; but I knew I couldn't walk back to the stables, as everyone would see failure clearly engraved on my face. But in that moment a woman's soft voice interrupted my anguish and despair, replacing it with hope.

"Can I help?"

All I could do was softly nod, hiding the emotional mess I'd become as best I could. Entering the round pen, she strode confidently towards me and inquired as to how she could help. I could do little more than wipe away my tears with the sleeve of my shirt and maintain my composure as she gently took the lunge rope from me and started working. No words were spoken, and when she passed back the reins she was just standing next to me, quietly encouraging and giving me back the confidence I'd lost.

And just like that... I was back on track. As I hopped on my bike to ride home later that afternoon, I looked back at my new painted "partner" then left with a reserved smile. That day I learned more than I taught the horse... and was old enough to put my bike away without offering it a carrot and a pat.

Most of my friends couldn't understand the days I spent in the pouring rain, ice cold winds, or oppressive heat I'd so willingly spend with the horses.

Well; there's this magic... A magic connection with a horse when it all comes together. This magic and invisible communication, trust, love and excitement that can only be felt by those who give their hearts to horsemanship. Magic I feel through their smell, their noises, movements and those first steps they take as they walk towards you. Horses are so authentic, so pure, so honest; unlike people. And it's for these and so many other reasons we connect with our horses and become horsemen.

But, I couldn't stay 17 forever, riding horses, hiding behind books and already I knew there was only one thing in life I was sure about. Horses were going to be part of my life, and I didn't want to end up in a normal job, slaving all week, dreaming all week of the weekends that would be over in an instant only to have to do it all over again come Monday morning. I saw people staying in relationships out of habits, not love, and I swore to myself I would never do the same.

So, despite my parents' concerns I may stay in my shy little shell for the rest of my life, I took a flight to Australia straight after my graduation. I wouldn't say I panicked; but I definitely ran, finding myself on a horse in Australia soon after. I worked droving cattle, and it was here I found unconditional love.

Jill was a terrible cattle dog, but one of the most beautiful souls anyone will ever meet. Jill changed me, and like so many amazing coincidences changed the path my life would follow. That little dog that follows my every move, followed me all the way back to Bavaria when my Australian visa ran out in nothing short of the most expensive decision I've ever made. I'd worked hard for it, but I'm proud my heart makes decisions for me.

But the next decision I made was out of respect for my parents whom I both love, admire, and care for dearly. They'd insisted I get some sort of trade of further education, and I was soon training as a vegan specialist chef in Berlin. My dreams could wait... not for ever, but for now.

Upon finishing my apprenticeship as a chef, my heart was once more calling Jill and I back to Australia. The smell of the eucalyptus, the cacophony of morning insect and bird song, the noise and feel of a leather girth cinching up before a mornings stock work and the dreams and plans for the new property my friend and mentor, Tanja, had just purchased were all just memories in the sweaty hustle and bustle of a restaurant kitchen. Memories... memories and dreams I wanted and needed to refresh, but getting back with Jill in tow was both expensive and difficult, so I decided to get as close to my destination via land routes as possible.

I had no idea as I boarded the train out of Germany heading for Mongolia, I would soon find myself in a world of adventure I would often pinch myself to believe is real. An adventure that saw me wander the wild Mongol steppe oblivious to the hazards with my ponies, my faithful dog and other travelers. But then I met Pete, and it all changed forever in an instant.

Chapter 2.
A leap of faith.

*"If your waiting to feel talented enough to make it,
you'll never make it."*
Criss Jami.

Sir Edmund Hillary is the greatest adventurer and inspiration for me personally. He drove a Massey Ferguson tractor to the south pole, and in 1953 he concurred the world's highest peak. He trained at Mount Cook "the most beautiful place on earth," and some 65 years later, every time I looked up at Mount Cook while hunting or took a New Zealand $5 note from my pocket, I felt inspired by his achievements.

He was not alone in his climb, beside him was a Sherpa by the name of Tenzing Norgay. Upon his decent he spoke those famous words heard the world over... "We knocked the bastard off!" But when asked who was the first up the mountain, Sir Ed just folded his arms and never said a word. This humble Kiwi legend knew it was Tenzing's achievement as much as his; and no act of humility has inspired me more than that of these two amazing men.

I never knew I'd meet Lu setting off alone, and I dare not compare us to these giants who's shadow we are barely worthy of standing in, but I am glad I have found a partner as strong as the one Tenzing found all those years before me. Who would have guessed she would have just been sitting there on a bench... waiting for me.

Every end is just a new beginning... right?

3 weeks pass; I want to cry. I want to scream. I close my eyes and can almost feel my hand running through the soft fur of a dusty stock horse. I can hear the birds; smell the eucalyptus, and see Jill by my side with a smile. It's all too much. I've failed. I'm stuck in Mongolia and the only way is back.

I'm angry with the Chinese border guards even though it's not their fault. I'm angry at myself. I'm angry with anyone crossing my path. Empty and frustrated, I still put on a smile... I don't want to explain, I don't want your advice, I just want to be left alone. My dream to get Jill and I all the way back to Australia failed due to red tape. Her vet passport had the wrong chip number written in it and I couldn't get her into China via the land borders; she had to fly in.

There was little more I could do other than sit in the park on a sunny day and feel miserable as I contemplated my inevitable return to Germany by train. It seemed so pointless, I'd come so far... we'd come so far and now I'd go back to work a kitchen job to save for a flight I couldn't afford. As if Jill could read my mind, her soft puppy dog eyes wander to mine as if to say, it'll be OK Mum.

"G'day mate, can I pat your dog? Seems a bit off seein' a Bluey here, what's ya story hun?"

A wiry blue-eyed stranger, "clearly Australian" holding a tall boy can of local beer strides confidently towards me, snapping me out of my gloom and straight back to the confusion I so often feel in places and moments like this.

"Sure, if she lets you."

I just watch him; ruffing up her fur, scratching her ears and tummy and giving her cuddles much to her delight. Without stopping, he turns towards me and talks as though he'd known me half his life.

"What brings you to Mongolia sweetie, and how the hell did you end up with an Aussie cattle dog?"

I wasn't overly warm in my reply, still trying to figure out this random fella's intentions.

"Well, we're trying to head back to Australia so I can get back to work on my friends horse farm."

He looks at me, this time for real. Clearly interested in more than Jill and more than just for the sake of talking to a woman. No; he looks at me because I speak his language, and share a love of horses, dogs, and living wild and free with fearless expression of individuality. He smiles; pulls up the sleeve of his shirt and a tattoo of a horse appears. His horse, "Digger", a wild Mongol pony from the far north of the country.

To be honest I didn't know what I should think of him at first. He'd given up on the society he'd left behind, and was here to just ride off into the sunset in the direction of Europe. Crazy... I shake my head; it's such a ridiculous thing to do. But I have just ridden around Mongolia for four weeks. He's fascinating and seems harmless, so I give him my number when he asks to chat later in regards to my horse contacts here. A line? He has horses... or genuine interest in me perhaps?

We part after a meeting lasting no more than 2 minutes... but I can't get him out of my mind. He was so direct and so different to most of the people I have met.

I am curious about this Aussie fella, so called him and we agree to meet up with him the following day. He came passed the backpackers I'd called home for months, and we walked Jill to a river flanked by tall summer grass and low sprawling willows and sat chatting over a few beers. He said I was crazy after my four-week adventure... laughing about the hazards I'd been blind to; inspired as much by my passion as I was of his. He talked with so much love; going on and on about his old dog who'd recently passed, and his pony that he "connected" with while wolf hunting for weeks on end in the forests along the Russian frontier. So much so that I immediately felt I could trust him. He was not only equipped for his ride, but prepared after years of training and life experience.

He fascinated, yet scared me a little in such a way I lost track of time and space... and then there's an offer. At first I think he is joking, but no, he is serious...

"You wanna join me? I'm going that way anyway, I've got no real plans other than living..."

I am speechless... Me? NO! Why not? NO! 10 Days later, hours and days of wondering, questioning... and I'm on my way with a happy scream!

The night before stepping into the abyss before me; I sat by a lake in Western Mongolia snuggled up to a man so soft and strong; so passionate and honest... I knew I was safe.

We sipped vodka and ate chocolate in silence as I pondered the future. I may have no clue about first aid, I don't know how to fish or to hunt, I don't know how to survive in the woods, I don't know what to expect, I don't know how I will cope with the upcoming winter and I don't know if I will be good enough. But I do know that the girl I was years ago, struggling with a paint in a round pen would be proud of who she would become; not because of her achievements yet. But because her heart hadn't changed one little bit; and it still chose the path she walked each and every day.

So there I was, gazing up at a billion stars hanging in space as infinite as the possibilities before me. I was at peace with my fate; as my heart had chosen it, not of the need to fit in, but because I wished to be nothing other than myself.

I've over thought everything my whole life, and I had no idea what to do... so I just went.

Ready as I'll ever be.

Although no pessimist, I view the world with scorn and distrust after a lifetime in physically and emotionally hostile environments. Almost poetically opposite to the childlike innocence and optimism of the woman I've had the honor of sharing thousands of kilometers on horseback with.

My background as a warrior, professional hunter, adventurer and outdoors-man meant I'd lived at very best, on the fringes of society. Just like Luisa, I was at a cross road in life, and in contrast to her feeling unprepared for the journey that lay ahead; my self-reliant and confident nature had convinced me I was capable of this undertaking. After all, I'd been working towards this sort of thing for my whole life and training specifically for the horse aspect of it for over a year.

But I know now, I'd be on a very different path without her. Every weakness she "felt" she had, she saw as a strength I possessed. But my obvious strengths often masked my many weaknesses and failures as a man, weaknesses I've since discovered Luisa doesn't have. These skill of hers are the reason both I, and the little dog are where we are today, and although it may sound dramatic... Luisa has saved both our lives a few times.

Years ago I remember a girl saying at a party "I'm so ready to have kids". She was 20, and there were a lot of older folk at the party that just roared laughing. A close mate of mine who'd had a few wines threw his arm around me and began one of those big brotherly sort of chats with me. He laughed at how all to often people think they are ready for children, but the truth of it is until you actually HAVE children, you really can't honestly have any idea what to expect.

The way you feel about your kid's is totally different to baby sitting someone else's, and the way you feel about your second child, then third is completely different again. He went on to say the first child is precious, the second is fragile; but the third one bounces.

He talked about how he had child proofed the house with kiddie locks on everything, and how careful he and his wife were with every single aspect of parenting; but 6 years later, he's just into it without a worry knowing what and how is needed and when. He made it clear, your never truly "ready" for kids, but as soon as you find yourself a parent, you quickly adapt and become one.

There are of course lots of things you can do to prepare before becoming a parent, and after finding out you've got nine months before a baby enters your life there's time to stock up on nappies and learn a bit more about what to expect while attending anti natal classes etc. But the thing is, you know its coming, its exciting and there's things to look forward too, shopping to be done and all sorts of new emotions as the big day approaches.

So how is any of this relevant? Is it even relevant? I've got to be honest... a crazy horse back journey is very much the same. I think it is as no matter how ready you think you are for a journey like the one Luisa and I took; your not ready at all. I met Luisa literally day one. I couldn't have predicted that in a million years I would have added her and Jill to the team within a week, and although I had pictured a dog joining me at some point, I never could have imagined an Aussie cattle dog!

But there were a ton of things I had done to prepare; and these are things that are worth a mention, although many of them again failed to prepare me in the ways I thought they would. My time in the military and as a hunter helped no end, but I knew very little about actual horsemanship as such. I started asking anyone I could for advice; and to be honest the response I got was not as I had expected.

Firstly, pretty much no one took me seriously, and I very quickly learned that the horse world is one where jealous bitchy behavior is stock standard. Further to this I think that not all, but many of those that have taken longer journeys or achieved excellence in horsemanship sort of see themselves as on some sort of pedestal. Even an internationally recognized group devoted to long riding "The Long Riders Guild", which I was told were a wealth of knowledge, took some 6 months to reply to me and when they did it was simply an apology saying their website was down. They have since sorted the issue and are now a fantastic contact to have.

Facebook has helped a bit, but again I wasn't taken seriously and had an endless stream of people running me down telling me I wasn't experienced enough and that what I was doing was stupid. So, I started buying books. But again, these didn't really help me a great deal and even the best of these were often pretty outdated. Borders and crossing them with horses has changed a lot since many were written, as well as the huge developments in technology for the travelers particularly in the form of smart phones and GPS.

The more I read, the more I found contradictory information and I came to realize the best way for me to approach the trip was to just go and figure it out myself. I would recommend others do the same, but time in the saddle before heading out was something I felt was the obvious next step. The people that can help you most are out there doing, and doing it every day for that matter. The cowboys, the horse breakers, farriers, race horse trainers and endurance racers. So, I went out there and I got into it; riding anywhere and everywhere I could with as wide a variety of horses and horsemen the world over.

Again, experiences varied greatly and once more I was exposed to the sad fact that much of the horse world is made up of could have been champions with tall poppy syndrome. There is exploitation, back stabbing and jealousy; as well as men and women that are both exceptionally talented and caring, offering endless encouragement.

In the year before I set off, I rode horses in 7 different countries, and with a huge variety of horsemen, and utter fools that loved imparting second hand YouTube video wisdom as their own.

I even spent a ridiculous amount of money, literally almost what Luisa and I have spent in the last full year on a Mongolian endurance race similar to the Mongol derby, which is regarded as the toughest horse race in the world. I did not compete in the Mongol Derby, and the race I did shall remain nameless as I felt it was an utter shambles from top to bottom. The reason I have chosen not name it is simple, many of the other riders had the time of their lives. I did not, I spent nearly $9,000USD on a week where I did things that were literally half as challenging or enjoyable as the weeks I'd spent in Mongolia prior to the race for less than $500USD.

Bitchiness, petty rivalry and terrible organization plagued the event which sells itself as creating a ton of jobs for Mongolians. Rubbish. They get paid cents on the dollar while the organizers keep the lions share. But; it was a brilliant experience none the less and I met some amazing riders who I constantly questioned throughout the event, learning each and every day a new trick to help manage a horse over long distances. For this reason alone, It was money very well spent and helped me no end in my preparations for the long ride I was now seriously planning for the following year.

The American riders in particular were an amazing wealth of knowledge, with many having competed in the Tevis Cup "a grueling 100mile race in the USA" as well as the Mongol derby. Just salt of the earth sorts; humble, knowledgeable and down to earth in total contrast to the organizers who treated us like naughty children, and at many times endangered riders through their carelessness and lack of proper preparation.

But for those new to adventure, and those that hadn't traveled much, it was the time of their life; and although I was constantly frustrated and ended up walking off at the completion of the final day, only to be disqualified for "leaving camp", I most defiantly found it a great way to learn in particular, the veterinary requirements for keeping a horse within its physical limits over longer distances.

The other thing I found very useful was the fact I was able to finally measure myself against the best in the world to see just where I stood physically and mentally when it came to my ability to ride a long distance. On day one of the 480km six-day race, I finished equal first with a man who is widely regarded as the finest endurance racer in the world. Six of us; four Americans and two of us representing New Zealand crossed the finish line holding hands in an act of sportsmanship I'd suggested as our group trotted together over the last couple of kilometers between us and the finish line.

There was one last water point where we stopped and agreed: we've ridden all day together; we cross the line together. The race director who was at the water point directed us to the finish line and we all set off, crossing the line at the trot holding hands while the photographer videoed us.

But; the track marker and race marshal unleashed upon us a hail of verbal abuse including enough foul language to make me, a former infantry machine gunner; cringe. Although directed by the race director to head straight in, we apparently hadn't gone around a tiny bit of pink tape some 100m away. It was a moment that set the tone of the event for me, and one that bonded us as riders.

I spent that night working alongside the doctor treating heat exhaustion in two other competitors, as well as supplying antibiotics and other medicine I carried that she didn't have on her. She was being paid less than $100USD for her 10 days as the race doctor. I left her with all of my medical supplies when I left, and it was sad to see how little she was valued in such a hostile environment. But at the end of day one, I felt as though I could stand shoulder to shoulder with the fellow competitors from around the world as a valuable and capable rider, as well as by shoulder to shoulder with the race medic.

But this race just wasn't suited to me, my attitude or my needs. Sure, I made the most of it; but this is a race designed for people who ride around the farm, do the odd cavalcade and want to have that "once in a lifetime" adventure horse ride that is fully organized for them. I came into it as a wild unshaven bushman who'd just ridden a 22-hour day hunting wolves the week before. 17 hours in the saddle, followed by 5 hours drunk, bareback with a slung AK47 in a mosquito filled willow swamp with only one tiny head lamp between 5 of us riding practically blind in the pitch darkness.

But I was learning valuable lessons, and a week after the race I was galloping a mare through the Georgian Caucasus's with a yearling and foal hooning along beside their mother and I. I didn't know it at the time, but that yearling would come to be named Pippa; by me, as I gently broke her to saddle some 18 moths later.

A few weeks later after a champagne breakfast in Siam Reap; I was cantering a stallion between the rice paddies of Cambodia, drinking warm Tiger cans and marveling at the ancient temples I visited on horse-back. It was here where I came to overcome my biggest fear of the upcoming ride: traffic. As I rode past a wedding blasting music, among a sea of kamikazes on scooters, past flashing signs, burning piles of rubbish and seas of people swarming about markets and street vendors I came to a very important understanding. If your calm, your horse is calm; with the correct rider and training you can desensitize a horse to traffic and for that matter almost anything.

I had a stint back in New Zealand working with race horses too, and it was here I really started to dislike thoroughbreds "TB's". They just aren't my cup of tea, but the time I spent working in the racing stables handling hot horses, tacking them up for the track riders and washing them down each morning taught me a great deal about a horse's body language. I went on to assist with breaking race horses at the same stables, and worked under the guidance of a master horseman and former rodeo champion who's simple and blunt instructions were as clear to me as the instruction he gave to the horses he worked with.

It was under his guidance I broke to saddle my first of now many horses. I would pay special attention to the work done by the farriers here too, and I'd always go and assist; watching, learning and listening to the advice given more often than not in general conversation, as we went about the daily jobs working with the breakers. I learned a lot from that wild rodeo champion; but again, I saw the bitchy underside of the horse world and the dramas that go along with working in an industry where opinions often vary so greatly.

Back in Australia, I was very lucky to spend a bit of time at a Warmblood stud farm in the Adelaide Hills. It was here I had one of my most serious falls, and it was in front of one of the most experienced riders I have ever had the pleasure of knowing. A genius in every sense of the word, she was not only a former Swedish model, world chess champion and horse breeder... she was an 80-year-old woman whom 17 hand stallions bowed in gentle submission too at a mere glance. I did the same.

She was amazing; fearless both on the ground and in the saddle, she told me something that I have never forgotten. I guess, when your 80 and have kicked arse your whole life, you're not afraid to tell people exactly how it is or mix words to save feelings. I was nursing my wounds, talking over beer in the stables when just straight out of the blue she walked over and said the most profound thing to me.

"Pete, some people should never be anywhere near horses. If they love horses, they should just get a post card with a pretty picture of a horse on it and look at that instead. No one has the right to just ride horses and some people just can't be taught. I can teach you. I just watched you make a mistake; you know what you did and you've learned from it. It was your fault and you know it.

A 500kg fresh off the track thoroughbred "OTTB" mare just landed on top of you, trampled you, smashed your face open and you're going to need to get that leg x rayed... and the very first thing you did was spring to your feet and stroke her forehead and calm her. You look at yourself, you don't blame the horse and you got back on, brought her home and groomed her before treating your own wounds. You got lucky, you could have been killed and it was your fault, but you still ride without fear. Well done."

That mare was a pig of a horse, it spooked at everything and hated a pair of lamas in a paddock near the stud farm where I was staying. I was helping out doing fencing and cutting wood etc and had use of the horses, in particular the rescue mare that tested me daily. I got her pretty good in the arena and in the jump paddock and would take her around the farm a fair bit, but rarely on the road as she was just a spooky mess.

I liked her though, I liked her because if I'm completely honest I was scared riding the Warmbloods, not because of their power, but because my favorite mare to ride was worth every cent of $50,000, and the grand prix level dressage stallion I rode in the arena was worth a lazy 1/4 million or so. I was just scared I'd step in a rabbit hole and break a leg or something, whereas the messy little OTTB would be lucky to go for $500 at the works, meaning I was riding with my mind in a better place, despite all her issues. That and I just learnt so much more dealing with a difficult horse, and my fall taught me more in 3 seconds than I'd learned in the months prior.

My accident happened as we approached the lama paddock, she pulled on the bit harder and harder, and I just had less and less I knew what to do in order to manage her behavior. She was an ex race horse wanting to bolt, starting to pig root and buck and I was managing as best I could to keep her under control without pulling to hard on the bit. It was useless, I simply wasn't good enough for a horse with the issues she had as a result of numerous factors, and as a result I learned a very hard lesson in the seconds that followed.

My mentor beside me had seen this coming I think... and had insisted I wear a chest protector and a helmet, two thing I am very reluctant to ever wear, but they most defiantly saved me serious injuries in what followed. I had her hard up on the bit, and I missed the warning signs... the last thing I remember was hearing

"DROP YOUR REINS!"

It was too late, I'd given the stroppy little mare nowhere to go but up, and she literally jumped completely upside down, landing flat on her back with me under her before I knew what hit me. She righted herself, standing on my left shin with a sickening crack I felt go up through me like an electric shock. I was on my feet as fast as her, somehow still holding the reins as I gently began calming and reassuring her.

I knew it was my fault, and in the months that followed I learned how to avoid situations like this; but in that moment I just stood dazed, nodding and saying...

"yeah, yeah I'm OK; please tell me I didn't hurt the horse! Is she OK?"

My mentor circled me on her horse, dominating my mare with her presence further helping to calm and manage the situation as she asked again and again about my well being. I placed my hand on the horn of my saddle to steady myself as a steady stream of blood from a split across the bridge of my nose started to fill my eyes. The smooth leather of my saddles horn was now jagged, and as I wiped the blood away to clear my vision, I was amazed to see pieces of the gravel road I had just landed on embedded in the soft leather. This horse had binned me totally and utterly. It was a real wake up call for me, and one I am very glad I got.

They say it takes 100 falls to make a cowboy, but it only takes one to wake you up to just how dangerous a fall can be.

But I wanted this. I thought of those Aussie blokes that fought the great war and how the legend of the light horse will live forever. How the light horse sported a plume of emu feathers in their slouch hats, plucked wherever possible from an emu at the gallop; a mark of a great Australian horseman. I've chased more than one poor bloody emu about on horseback since, and as I sat with a bag of ice on my shin and a cold Victoria bitter in my hands, the sun sunk behind the gums. Pink galahs and cockatoos all screaming their heads off, Pobblebonk frogs sung in the dam below the sleep out in which I was staying as I stared at the fire dancing in the brassier in front of me. I contemplated the exact moment I knew I wanted to be a horseman.

I was a boy, and it was probably sometime in the late 80's. I was watching a movie called "The light horsemen". On October 31st 1917, a 3-mile charge of the 4th and 12th light horse, a mere 800 light horsemen charged against artillery, machine guns and some 4000 infantry to seize the town of Beersheba; a town named for its 7 wells. Wells that must be captured to avoid an absolute military catastrophe.

In a scene moments before the charge, the diggers sit with their horse's. Horses that hadn't drunk in 48 hours; and one digger mutters "bugger this", before giving his horse the last of his water poured into his slouch hat.

That's what I wanted... I knew I wanted to experience that sort of bond with a horse; the kind where you were in it together, and you'd give him the last of your water, as he would give you the last of his strength in the coming charge.

I knew I had a long way to go, I knew I had a lot to learn, but as my mentor and my fall had taught me that day; I had no fear of what lay ahead and would get up no matter how many times I got thrown. This is the best preparation I could have been doing for an adventure, sitting there grimacing through the pain, sinking a few cold ones dreaming. Dreaming and inspired not by those with great YouTube channels full of advice, not by books, not by Facebook groups or paid riding lessons.

But inspired by the blokes from the bush like me that went into a war with no idea what to expect; blokes that were handed a horse that had often been mustered out of the Blue Mountains and green broke at best. Blokes that became legends; and their horse's like the famous "Bill the Bastard", who too became legends in their own right.

Having empathy, a willingness to learn, no fear of failure and a genuine lust for adventure... that was the best preparation; and I now understood my mentor's words when it came to who should be around horses, and who should get a pretty postcard. I belong with horses; and although unskilled, inexperienced and with no idea what lay ahead...

I was ready.

Chapter 3.
Sunny days.

"A willing partnership starts with kind hands"
Lorrie Duff.

As I already mentioned, this isn't going to be one of those books where we tell the tale chronologically, day by day, and talk about the places we visit. We want to talk about how we feel now, and how these experiences made us feel at the time. Sure; were going to share some of the more memorable moments, but our overall goal is to inspire others to take a journey of their own, in their own way, and we hope prepare a little better than we did.

Luisa and I are both a bit strange, and I think it's important to say right from the start that we were both looking for some sort of romance with hardship, while sharing kinship with other living things. A way of showing and feeling empathy with all living things; even those that hunt us, and those we harvest to sustain our own lives.

It's about connection with the very earth itself, and others that walk it beside us, and those that have walked these paths before us. Horses are the means to connect in a way other means of transport cannot; as they force upon you needs that other transport doesn't.

When you travel by horse, you must interact with the environment in such a way, that things you would simply glance at from behind the glass of a speeding vehicle, become essential for your very survival. There is a closeness and connection between all living things those that walk paved streets in soft shoes cannot feel and for us, that urge to walk bare foot in the woods is what drew Luisa and I together; but our Journey began hitching a ride in a little Toyota wagon named "Penelope", with a man named Gunter from Aachen, Germany.

Living the dream.

Why has our trip worked? Well, in short it has worked for the reason many others fail. Flexibility.

We're just focusing on living and as I type Luisa's is on the phone to a guy in Turkey trying organize horses for the next stage of our ride. Can we get them into Europe? We think so, were still trying to figure that out... maybe with winter coming soon we will catch the train to Greece and continue the ride from there around the Mediterranean until the passes open and we can find our way up into Europe by any number of routes that may present themselves as a result of whatever adventure lays ahead of us. And it's been this rigid inflexibility in the maintenance of flexibility that has seen this not become a lonely solo horse trek into the wilderness, but a love affair with life and from that Luisa and I by complete accident fell deeply in love with one another.

That flexibility began in Mongolia. we'd both ridden there, and although it would have been great to cross once more on horses, visa's and the coming winter had us hitching a ride in a little Toyota wagon piled high with gear, and its roof covered in spare tires.

I woke beside the car, half in my sleeping bag and covered in sand long after the sun had risen. Beside me, a gorgeous young German girl was doing yoga in her underwear, or as Luisa would say in her broken English, "making yoga".

We were by a lake on the Mongol steppe, with nothing but an endless expanse of low rolling grass land in every direction as far as the eye can see surrounding us. Our driver, and now good mate Gunter wandered the shoreline some 50 meters away with his camera, taking snaps of birds and the cattle and spartan loosely scattered across the stunning spartan landscape. Foggy memories of the night before return; Luisa singing and dancing about the fire, vodka, laughter and at some point, skinny dipping by moonlight in the lake which seemed surprisingly warm for this time of year.

The trip was off to a great start, despite us having been bogged a good dozen times getting to the lake; oh, and the fact that the gearbox had dropped its sump plug the day before. With all the fluid having poured free, Gunter decided we would risk it; I however felt this probably wasn't the best course of action, and with the help of a random guy that just wandered out of the steppe on horseback and a friendly passing motorist, I'd carved and hammered a stick into the sump plug hole as the missing bolt was somewhere between here and Ulaanbaatar.

We filled the gearbox with engine oil, and shared a bottle of roadside vodka with our new friends as we worked, laughing at how the Mongols, just like the Aussie under the car agreed you can fix pretty much anything with the right attitude. Our attitudes and the vodka must have inspired Gunter, as he soon applied that "she'll be right" attitude to the rest of the afternoon, deciding we would head for the lake, rather than into town to fix the gearbox.

Luisa and I slowly stirred into life; her with yoga, but I opted to crack a beer and have a bit of a scratch. We soon both stood half-dressed by a shimmering lake bathed in warm autumn sunshine in the very heart of Mongolia. Yep, trips off to a pretty bloody good start if you ask me. Gunter returned seeing we were both up; soon lashings of thick local butter covered bread, crudely cut with a pocket knife, were being stuffed in our faces alongside wild berry jams, wild honey, and sliced salami.

We laughed, contemplated fishing or moving on, but for the most part we just enjoyed soaking it all in. Jill was sitting by my side making her best "can I have some food please Pete" face, and Luisa was bubbling away as she so often does about how much she loves Christmas, when a lone shepherd on horseback approached us, hobbling his horse no more than 30 meters away. Gunter got excited, grabbing a spare cup of coffee and heading straight over after exclaiming to Luisa and I...

"I'm off to buy us a goat"

Lu and I looked at each other, and there was a long pause as smiles slowly spread across our faces before we burst out laughing. We'd kind of joked about it before, and there was clearly no issue with me knowing how to kill, cook and clean a goat; but just the matter of fact way in which he went into it was just one of the many things we loved about Gunter. We soon joined in the conversation, and the shepherd was more than happy to sell us a goat for $15USD... but we'd need to catch it first.

Gunter and Luisa decided to film and take photos, though in the end neither captured the hilarity of what followed; as the herders crafty goats dodged his lasso twice before I opted to just run and jump at the mob, catching one by the foot in the sandy bank of the lakes shore with the style and grace of a drunk falling down stairs. The Mongol laughed, but was still none the less impressed at my commitment and was soon happily on his way on his horse, as we bundled the goat into the car before heading on towards town.

There was no point killing the goat just yet, goats are funny sorts of things and it wasn't particularly fussed as it road towards town with us. But as it was a warm day, we feared it may spoil and its always easier to cleanly butcher an animal after its fasted for a little while. Upon getting into the next town, we managed to find a bolt that not only plugged the sump correctly, but unbelievably allowed that little Toyota to make it all the way back to Germany without another hic up! We stocked up on food, I did a few repairs on the car and then once more we were off in search of nothing more than where the day may take us; and that day it took us down roads no sane person would take in anything short of a good four-wheel drive.

But we didn't have a sane person behind the wheel, we had Gunter, a goat sitting in the backseat, Jill in the front and Luisa and I sitting in spare tires on the roof drinking beers. The "roads" quickly deteriorated into a maze of dusty cattle tracks criss-crossing the steppe, interrupted by the odd low rocky outcrop as we cruised along, rarely exceeding 40khp. It was gorgeous. Sunny, warm and the scenery was far from empty; with mobs of wild horses, yaks, camels, sheep, goats and the odd random fella on horseback looking puzzles as a car chocked full of luggage "and a goat" cruised past with two friendly strangers waving from its roof.

If ever there were a day for the saying "the goal isn't the destination, it's enjoying the journey", then this was it, and Gunter certainly asked far more of that little car of his than it was ever designed for! At one point both Jill and the goat joined us on the roof for a photo, and I will add a very careful disclaimer in here that at no point was that goat treated with cruelty or contempt. Bear in mind, Luisa is a former vegan and vegan trained chef, and neither of us believe in allowing anything to endure unnecessary suffering.

It was a special day, and as sunset approached, we looked for the perfect location for the night's campsite, settling on a secluded and sheltered stretch of grass between two beautiful towering rock features. They were deep red ochre in color, in total contrast to the soft brown shades of the surrounding flat, semi-arid steppe, and as the sun fell their colors changed in a stunning display of breathtaking natural beauty. The goat was sacrificed for the nights feast with both respect, and efficiency, and I was surprised to see the former vegan so curious and helpful during the whole procedure.

I would grow more and more to respect, then love, this woman in the coming year, but I'd be lying if I said this wasn't the point where I recognized she was special. Not because she watched me kill a goat, but because like me she longed to feel connected with and understand every aspect of the world in which she lived. She wanted to be present at the harvesting of the meal, during its processing, to help build the fire on which we would cook it before telling stories under the stars and sleep on the earth itself. I knew I had met someone very unique, and was as happy as I had ever been.

The meal was a feast, and as we ate in camp, we could hear the howl's and chatter of the wolves above us where our little sacrifice had occurred on the ridge at sunset. We had left them an offering, of course, one they were now seeming to devour with much glee. Jill, never the bravest dog huddled close to Luisa, as we all bathed in the warmth of our glowing fire by that little Toyota on the steppe under that sky... that amazing star filled sky you can't even believe is real until you gaze up at it from a place like this free from light pollution.

You feel infinitely small, but connected; almost as though somehow all those tens of thousands of generations that gazed up at the stars before the modern technological revolutions, just wondering what the future would hold... are right here with you. Luisa and I both knew, this was the beginning of a journey that would see us unafraid when our times come to pass on from this world, knowing we had lived, loved and laughed; not just had the words written on a fridge magnet, calendar, painting or photo frame.

Heads may differ, so long as hearts agree.

Day one of our ride, we lost Willow at sunset. I glanced up to see him heading for the distant village across the low rolling fields waist high in Autumn grasses and flowers in a lazy bend of a wide river some 30 kilometers out of Pavlodar, Kazakhstan. We'd just bought our first three horses, Bill the Bastard, a tall powerful buckskin stallion, Jac, a gorgeous chestnut gelding and stock horse, and Willow. Willow was a cheeky grey gelding, and I do mean cheeky!

The other two were grazing quietly on their tethers and to be fair, this was a procedure we were both yet to master as neither of us were confident enough hobbling horses at this early stage. Luisa saw our beautiful full-size pony we had planned to use as a pack horse sauntering off seconds after me, and the two of us shared the first glance in which no words were spoken, but a full conversation was had. As I sprinted to head off Willow on foot, knowing full well I had no chance of catching him without Luisa's help, I wondered optimistically if she and I were on the same page mentally.

I got ahead of Willow rather easily, and began to shepherd him away from his intended destination of his familiar home some 15 kilometers away. He eyeballed me skeptically, dragging his lead rope and tether behind him letting me get close; but never close enough... and then I saw her. Luisa cantered Jac bare back over a low rise with the setting sun behind her. Her dark silhouette on a glowing little gelding a stark contrast to the brilliant orange sunset behind her. They moved as one, and breaking to the gallop I was momentarily stopped dead in my tracks.

I'd know Luisa for weeks, we'd even shared a kiss, but until that moment I hadn't even really noticed her as someone I may find myself having a romantic interest in. Still somewhat dumbfounded by her grace, her poise and total and utter composure riding that beautiful chestnut at sunset I stepped on willows lead rope with a foot, a gathered up the runaway with Luisa having confidently predicted and blocked his movements on what was clearly a capable stock horse. We didn't say a thing, she just smiled, and after a quick pat and a chat, I grabbed a handful of mane and threw myself up onto Willow to ride him home beside my partner.

Luisa and I were pretty much married at first sight. Days after our first meeting we were heading off on an adventure in many places where a man and a woman traveling together is only really socially accepted if you're are in fact a married couple. So, we got rings and that was that. We would travel as a married couple from the start, as it would save a lot of explanation and further to this, hopefully offer Luisa a feeling of added security. We got on from the moment we met, and as there is a substantial age difference between us, just both sort of felt we'd be friend's and that would be that. Neither of us had any interest in romance, getting tied down or for that matter even considering anything more than a ride and a great adventure with a bit of company.

In a way, we both went into our relationship selfishly, and certainly with no expectations of one another. It was perfect really, and set us up for what would equate to nothing short of a cheesy Taylor Swift song; you know, the type where the fella realizes he's in love with the girl next door.Well, for us those first couple of weeks with Gunter took us from two travelers, too two travelers entering into a partnership, and each and every day knowing neither had a romantic interest in the other we showed each other nothing but ourselves.

I think for us both, this is where something special began to happen, but to be honest, I hardly even looked at Luisa as a girl. She is "as rough as guts" in the correct Australia terminology, with a shaved head for easy maintenance, no girly nails, no makeup at all and isn't exactly girly. A sports bra and a singlet over jeans was as racy as it ever got with Luisa, and I wasn't any better either, clad in my usual mix of Sitka and Swazi outdoor gear.

The only way to describe our outward appearances at a glance would be functional and genuine, after all, this was an adventure; not love island. But the odd hug became hugging, and on those last freezing nights before we crossed into Russia, we sort of had a moment.

So why is this in the book? Well, I am writing about it simply because it's what happened on the adventure, and these sorts of things happen in the strangest of ways and in my experience most often when you're least looking for them. Tinder, Bumble etc; that's how most people hook up these days right? And how often do people find what it is they are truly seeking?

Well, we were fed up with the instant gratification and entitled generations of consumers telling everyone how they should think and feel so much so, that the two of us just found ourselves sitting on the roof of the little Toyota together sipping vodka, eating chocolate while looking at the stars together chatting. It was freezing, but we were bathed in the light of a full moon, and Luisa, who is a cuddly affectionate sort, was all snuggled up with me... kissed me as we were making a selfie video together. Um... righto. That was unexpected, and the strange thing was it felt somewhere between kissing a friend, and... I don't know? It wasn't exciting physically, it just sort of; I don't know... felt right?

Well, nothing was said, and from here we just continued as per usual and shared a room whenever we found ourselves in a guest house or hotel. Not once did either of us try a thing, nor even consider it as we were just two mates who had become very comfortable with each other almost... look this is really hard to describe. She was just, you know... Lu! We did our own things and had our own thoughts on how things should be done, but were going in the same direction, so ran with it.

We were just a pair of birds perched on the same branch singing together, rather than a pair caged, or flying past one another. It was the strangest thing really... We were both just so focused on the adventure we never really stopped to notice each other, seeing that "real side" of one another over the "the person you pretend to be when seeking a romantic connection."

It was over four months before Luisa and I had a night apart, and after a full year together we've traveled an 8th of the worlds circumference together since on horse, boats, trains and hitching. We've ridden over 5,000 km each on horses, and spent around 2000 hours on or around them during this time, and by choosing not to fall into a relationship of convenience or conform, we accidentally fell in love. I didn't even notice it happening until Lu started telling me she loved me, more in the way she would say it to a sibling before hanging up the phone... but it was genuine.

Things very slowly grew from here, and I'm honestly struggling to put our feelings into words. We are partners in every sense of the word after having seen no matter how tough things have got, as long as Lu, Jill and I are together; the sun is always shining when we close our eyes and hug.

Upper body ripped and a cricket smile.

As the sun rises, it bathes everything around me in golden hue. I haven't opened my eyes yet, but the light flooding through the tent is hard to ignore. I know that a certain little cattle dog is staring at me, waiting for my eyelids to lift, my body to move. Even though Pete had already let here out, just like every morning she is still waiting for me to raise so she can tell me about the wet grass outside the tent she's just bounded through, the smells of the jackals, and that she had bravely watched over us all night from under the sleeping bags and deserved a hearty breakfast as a result.

I can't help it but smile as I open an eye a sliver; straight away her tail starts slowly moving before going full speed as I open my eyes all the way. I great her in the usual manner...

"Morning beautiful little girl"

She bounds towards me and I reach out to hug her, but she's straight into my sleeping bag, back to the warmth and the intimacy with a few happy grumbles and groans. She's a bit wet and cold after her morning walk, and as she presses against me for warmth, I feel every muscle relax as she lets out a happy sigh.

The noise of the horses grazing softly creeps into my ear's as my brain slowly stumbles into clear thought as last night's dreams fade to foggy memories. I listen to Pete's soft voice, not more than a whisper, talking to Bill the Bastard. Sitting up, I watch the steam slowly rising from the pot on the fire as the smell of hot chocolate and willow smoke drifted into the open tent. I should really get up and start the day; but instead, I fall back into bed. For the next few minutes I just soak in all the peace and quiet, excited for my day to begin with a moccachino I know Pete will have dropped at least one whole bar of chocolate into. I dose off almost immediately, and am woken by Pete...

"Good morning Princess."

His hands are weathered, dry and covered in little scars but his touch is soft as he gently strokes my head. Not unlike a cat, I stretch to get more, to make this moment last just this little bit longer before I sit up for my hot brew. Pete gets back to the horses while I slurp the warm, sweet muddy water like mix of chocolate and coffee with fresh milk sitting in the door of the tent. The excitement of the day to come raises just like the sun before me, but I need to pee, so kick Jill out and try to improve my skills getting dressed in the sleeping bag without letting cold air creeping in.

Jac is mine today, as Pete and I switch horses most days, and while I walk towards him happily bubbling away... I am received with a vicious look, but pay no attention to it. I greet him as usual, "Hey cranky man" and begin stroking his forehead. He's a gelding, but dammed if he doesn't act like a bitchy mare most days. He softens under my touch, and we're soon getting ready to go. I still struggle with the saddlebags, but Pete is already done. I pretend not to look towards him, as we have a sort of "unspoken race" saddling our horses each morning that I may have lost... again; but couldn't stop smiling about.

For those that can't or don't ride... It's hard to describe the feeling of cinching up each morning. That magic feeling inside you heart and soul as you grip the horn of the saddle, slip you left foot into the stirrup and swing your right leg over onto the back of your partner.

Minutes later, your melting into the movement of the horse, and as the hours pass by, its as though time stands still. And today we were following a power line, the connection between the numerous small towns of the Kazakh steppe. Over the last few days, Pete had me almost convinced that trolls existed and that the power lines were used to keep them in their territories. He pointed out the remains of burnt and shredded tires on the side of roads and tracks, telling me they were offerings for local trolls and they had been fed on when it looked as though chunks were missing.

I smiled a lot, so did Pete, and he was always trying to introduce me to his beloved Australian Hip Hop. I understand why he loves it; its never about gangster crap, it's honest and uplifting, about love, growing from your mistakes and never giving up in the face of adversity.

Urthboy; a father, a rapper, and gentle soul's song "The Arrow" resonates over the whole steppe around us, blasting from the Bluetooth speaker in one of Pete's saddles cup holder's. The lyrics are particularly touching, and I begin to understand why Pete refers to it as poetry over music...

"My arms stretched either side... And who knows where it may end... The weird thing is that I've never had a better f#$&*ng time in my life..."

I glance over at him; his arm's wide open stretched either side, his lips forming the words.

A smile instantly spreads across my face, not the one most people see, but my real smile which is slightly crooked; a fact Pete loves reminding me of for his own amusement. I sent him a text message about it once, but used the word "cricket" by mistake, so it's now known as my cricket smile. Pete is best described as "wiry", and despite being very powerful there's not much too him, so he often refers to himself as "upper body ripped". It's strange, having been around Australians for most of my exposure to the English language, giving those you care most about a hard time has just sort of become a habit now.

I drop my already loose reins, and we both sing out loud; with my best German attempt at English words I don't understand hilariously mixing with Pete's Australian. As if mocking us, our boys we sit atop throw each other a glance, and I'm sure if horses could smile, they would have, laughing at the clowns singing and dancing about on top of them. Jill's eyes caught mine at some point too; excited, looking at me as if asking what fly are you chasing and yelling at mum!

As the last notes disappear, our laughter follows like an echo, only to be swallowed by the endless steppe surrounding us. The horses haven't even changed their pace, straight forward on as if on a mission, they look after us just as we look after them. Light clouds hang above us motionless, almost reminding me of a dance floor filled with people moving as one to music yet somehow frozen in time; the sun watching over us from its perch in the Autumn sky, the rhythm of the horse beneath me, Jill, and Pete right by my side... there is no place on earth I would rather be.

But my grumbling stomach interrupts my romantic thoughts, with a sticky sweet Snickers bar the only breakfast I'd had other than my hot chocolate and coffee mix from my painted tin mug. I love that mug; it's been with me some time now, and its hand painted farm animals always remind me of my time mustering in Australia.

Pete doesn't carry a mug, he carries a small silver bowl gifted to him by a Mongolian; it's carved intricately in Chinese designs, but has a Mongolian symbol on its underside. It was stolen from the Chinese when the Mongols sacked their lands hundreds of years before, and was given to him the day after he visited the resting place of the Great Kahn and paid his respects. Man, a girl's mind wanders on the steppe and even more so when trying to write about it.

My eyes drift lazily about us, trying to spot a place for lunch and no sooner had those first grumbles from my stomach began, a promising spot opened up to our left, and a simple look to Pete approves what I already had decided. My energy shifts from that lazy Sunday morning couch potato, too Friday afternoon excitement.

The grass is good, and as we pull up the horses, before I can even come to a full stop Jac's nose already disappears into the feed beneath him. The well-rehearsed routine of unsaddling doesn't take long, the hobbles are on in no time and the horses enjoy their freedom "with boundaries" knowing full well to make the best of it.

Minutes later the JetBoil has done its job, and the water's boiling. 2-minute noodles and a can of horse meat meet crudely in a pot, are stirred and left to stew a while as Pete's silver bowl, half filled with vodka is handed to me by those caring, battle scared and broken hands. It's strange, the Pete I know is so soft, so full of kindness and empathy, particularly with Billy, that its hard to believe he has such a painful past.

While our food does its cooking, I stretch my legs before finally sinking down into the saddle blankets laid out just like a comfort of a couch. I don't plan to move for the next 40 min, closing my eyes and turning my face to the warmth of the sun.

My nose starts tingling, I pull my head back, opening my eyelids to be surprised by a white flower; maybe the last daisy Pete will find for the season and a warm bowl of stew. Rather than resting his body, Pete has used his time before lunch to find this treasure for me. Autumn has destroyed all but the last of the Spring and Summers flower fields, and only the hardiest flowers have kept their heads up high. This itself gives my flower a beauty all of its own, one hard to compare to any other in this world.

My whole body feels like exploding; like my hearts too small for all the love and happiness overwhelming me. As the first warm and salty stew hits my stomach, I stop to capture this moment; I soak in the light, the peace and quietness, the noise of the grazing horses, the feeling of Jill pressed against my leg, trying to steal the blanket from under me and my soulmate passing over more food. I ate in Michelin star restaurants throughout my apprenticeship, and had never even touched 2 minute noodles before meeting Pete. I'd grown up with the homemade organic cooking of my mom, and was always looking for new exotic combinations, the crazier the better, yet here I am with the most simple meal feeling like a queen.

Without warning the horses suddenly stop grazing; staring towards a stand-off trees low on the nearby hills. Pete's stopped eating too; he's staring in the same direction as the horses as he drops his spoon and quickly throws a glance at his Mongolian horse bow and arrows sitting among our gear. A shiver runs up my spine, we can feel the eyes on us just as the horses can; wolves most likely or maybe a snow leopard, though I doubt one would be so bold as to bother us. We sit still for a while, trying to make out the source of unseen predators as the horses decide grazing closer to us is their best option for the rest of our lunch break. Once within only meters of us, the boys put down their heads to keep eating, and after looking at each other, we do the same with the last bit in the pot going to Jill.

I place my head on Pete's lap, soaking more of this moment, this place... this feeling.

The magic of this trip isn't about the moments that seem special because of the landscape. The magic for me is its honest simplicity; and sharing it with the horses, a dog and a man who I had only drunkenly kissed once by mistake, but was falling in love with by the minute. Daily life... I loved it!

Just another day.

Some day on the step were just... perfect. For me none moved me more than a day where there was a real chance of seeing a snow leopard. We didn't of course, but just the mere thought that we had ridden into a place so wild at any minute we could disturb a sleeping leopard, spooking it from its resting place as it snoozed among the granite boulders high above the lakes we had left behind us that morning was a very strange and special feeling. I mean, a snow leopard! One of the world's most elusive and mystical creatures and here we were, a couple of clowns on horse-back with saddle bags patched and sew together after several small disasters and one huge one, bumbling along the steppe together is search of... well truth be told neither of us had figured that out at all yet and all we knew is we felt great.

We had been riding for a few weeks now, and our boys were quiet literally pets as much as they were horses. Luisa and I rode beside one another in the bright Autumn sunshine with huge smiles, our reins draped carelessly across the horse's necks as we shared drinks, snacks and endless laughs mooching through the forested hills of a visually breathtaking Kazakh national park. Lunch was particularly enjoyable, sprawled out in the birch tree thickets on the crest of a ridge in the blazing sun of a warm still day. It was heaven, even if we were only eating our usual noodles and canned horse meat, but I had a surprise for Luisa that night... I'd scored some meat and fresh vegetables in a small store that morning and now all I needed was an excuse to get rid of her at dusk, to surprise her with a romantic dinner in a location I was still yet to find.

To be honest, that was one of the most romantic parts of the adventure, never knowing what was ahead and even if we would find a suitable camp, or make do. But for now, I lay sprawled in the sun with Lu and Jill, both of whom had fallen sound asleep after the mornings ride. Jill lay curled in a cute little ball beside Luisa who was slumped awkwardly into her saddle as if it were an arm chair of sorts, making cute little snores and snuffles much like her dog beside her.

I got up, wandering towards my beloved stallion Bill with a brush in hand. Hobbled, but still able to walk with relative ease, he began making his way over to me as fast as possible upon seeing the brush. Bill loved being groomed; but even more so, loved grooming me as I groomed him. As I stroked his neck and back, he would bend his neck to me and softly nibble my back and shoulders in the softest and kindest way a horse of his stature could do so. Despite his power, size and at times attitude, he really was my boy; and proved time and time again at heart he was just a big soft loving marsh mellow on the inside.

Bill and I chatted for a bit; before Lu surprised me by slipping her arms around my waist and resting her head on my shoulders as Billy and I continued to chat and groom each other. We'd had a good break of over an hour, and both agreed without a single spoken word it was time to push on another 15 kilometers or so before calling it a day.

As always, suitable grass, water and a safe campsite dictated when and where we would be stopping for the evening, and the more time we had up our sleeve to find this, the more likely we were to have an enjoyable night. As we cinched up, I looked over to see Luisa's "cricket smile" having beaten me by a nose in our unspoken race to be the first saddled. I laughed, playing my loss of as due to spending greater care grooming and bonding my boy to Luisa's victory scoffs. But who cares at the end of the day, it was all in fun, and we headed off once more.

The movement of the boys beneath us, and Jill's constant bounding beside us while looking up at Luisa and I in need of constant approval and love just had it all fitting together perfectly for us. As we cleared the forest, before us lay a good dozen kilometers of open rolling pasture falling away beneath us with a small forested ridge in the distance that would do perfectly for this evening. It didn't take long to cover the ground, and we were soon pushing through birch, poplar and stone pines into a rocky granite ridge covered in low scrub looking for a perfect camp site.

Well; we found one that literally took my breath away, and as I dumped my saddle, hobbled billy and helped Luisa drop all her gear I knew I'd had time to set up the perfect romantic evening as Luisa had decided to gallop to a distant village that would most likely hold a store.

She had decided to do a quick beer run on Jac while I took care of camp. Lu sauntered off at the collected canter through the trees, ducking and weaving her way down the hill before breaking to gallop the open steppe falling away beneath our ridge. I knew I didn't have too long to get her surprise ready, as although a 10km round trip, that girl could really move on a horse when she wanted too and with an hour or so before sunset, I knew she would hustle.

I had the tent up like lightning, as Jill sat in defiant sulking protest to having been left behind by mum. Jill and I were cool; but I was nothing to her compared to Luisa, the two were just completely pair bonded and Jill just sat patiently waiting for Lu's return. As Bill grazed about, he too looked for his mate Jac, occasionally calling out while they both stared towards the trees, Lu and Jac had disappeared through minutes earlier.

I set about quickly gathering wood, literally dragging an entire fallen pine tree to the camp site with the help of Bill "fully planning to tell Luisa a troll had thrown it at me". I soon had a solid start on a fire, as I prepared a foil wrapped feast of herb and onion meat loaf with roast potatoes and a red wine I'd tucked away for a while ago in secret. Luisa may be a chef, but campfire cooking is something I've been mastering since cub scouts. I was laying out our bed spaces in the last of the days light, when Jill's frantically wagging tail combined with Bill's excited calls announced Luisa's return.

She casually trotted back in along the narrow trail with a beer in hand and a "cricket smile" across her face. She casually greeted me with little more than a "hi", before giving the ecstatic little dog the love she had been so desperately waiting for.

Amy Sharks "Love Monster" album playing as Lu groomed Jac after ditching his saddle, the smell of the campfire and cooking meal, the setting sun at the end of a superb day's ride. It was a prehistoric tapestry woven just for us, perfect, and almost overwhelming in its grandeur.

The meal was a stunner too, cooked to perfection and the sneaky bottle of wine I'd saved just topped off a day we will both remember for the rest of our lives. It was moments like this, or even just the thought they may exist that brought us both to the steppe. But having someone else to share it with, the dog between us, and the horses grazing nearby just summed up for me... everything the world had forgotten what being happy is really about.

Chapter 4.
It takes 100 falls to make a cowboy.

"Anyone can give up, it's the easiest thing in the world to do. But to hold it together when everyone else would understand if you fell apart, that's true strength."
Chris Bradford.

Believe it or not, Luisa has only come off a horse once during the entire trip, and that was only because she became hopelessly tangled by a panicking pack horse which literally dragged her from her saddle. I've come off once thanks to Jill's misplaced enthusiasm to "help", but fortunately my head stopped my body from getting really hurt on that fall. Another time equipment failed and I broke a bone in my hand, and I avoided death by the narrowest of margins by sheer luck and the presence of such a strong partner on more than few other occasions.

But we experienced loss we were not prepared for several times over, and on one occasion almost lost a very close friend we care for a great deal in a tragic accident. The grey ghosts "wolves" took from us, our mistakes cost us, illness took from us and accidents rattled us to our very cores...but we got right back on.

Dust yourself off when the wind's knocked clean out of you and wipe those tears when you put your hat back on, blaming dust in your eyes if you're crying; there will be time for tears later... right now there's cowboy stuff needen doin', and this ain't no time to be belly achin! Cowboy up, crack on.

What now?

Sweat runs down my neck. My throat is dry and screams for water. Fear overcomes me... every step feels more uncertain than the last. Jill, wagging her tail as usual, has no idea what's just happened. Walking straight ahead, I look towards how I can only imagine a battlefield appear from Pete's rare and obscure occasional references.

The village to my left seems like a ghost town; no one to be seen, not a sound to be heard. The sun was so bright on that perfectly day, blue sky above, warm and with a gentle breeze barely moving the trees. The cemetery on my right looked on with scorn, like a reminder I should be happy to be alive after what had just happened. My eyes follow Jill, she seems to try to figure out if all of the scattered and torn belongings are ours, walking from one to the other, excitedly sniffing. The contents of our bags are spread all over the ground, mixing with the first leaves of Autumn now peaceful after the carnage of the accident some minutes ago.

Is this it? Is this where it all ends three days in before it's even really begun?

Pete's saddle lies on its side, lonely almost; a few meters away in what feels more like a scene in the movie than real life. I am lost, overwhelmed by my feelings of guilt, frustration and helplessness...

I should have stopped the pack horse when I noticed the piece of tumble weed carried by the wind. We'd only had our 3 horses only for 2 full days, and we were still developing trust. My arm hurts, ugly bruises were starting to appear on my skinned, tanned over a long summer of adventure and failure on the Asian steppe. The event comes back to me in pictures, grim and painful; our pack horse Willow panicking as a tumble weed blew towards him, he broke to a terrified gallop cutting circles around Jac and I. I did all I could to turn with him, desperate to calm him with words and just as he seemed to reduce his speed a little, a bag tore... and it was all over.

Willow utterly lost it, equipment went flying in all directions. Jac, who had remained composed through it all stumbled and hopelessly tangled in the lead rope, I was dragged from the saddle, slamming into the ground with a sickening thud that knocked the wind from me as Pete desperately managed and turned Bill.

Reaching the spot where it had all happened moments earlier, I pause for a minute, wipe my tears and compose myself before picking up piece by piece our lives... everything that is us, piling it at the side of the road by the cemetery gates.

The clatter of hooves makes me look up... Bill! Relief overcomes me, Bill the Bastard was back! Led by a cyclist, the Bastard comes closer and closer, our stunning buckskin stallion trotting through the fallen leaves bathed in sunlight looking as though nothing had happened. I walk towards them, stopping right in front, as I try and find the words to thank the young man as he passes over Bills lead rope. He waves, bows his head and turns around, and is gone faster than I can blink. Now calm as a lamb, Bill follows me towards the neatly stacked pile by the fence where tie him up and hobble him as if to say

"you're not getting away again."

I mumble to myself, and hadn't seen Pete since the incident; I knew he had been hurt badly, but he was off so fast after the horses I just stayed put with the gear as he'd asked, knowing he would be back, hopefully with at least one of our three horses. I sent him a WhatsApp message, a picture of Bill tied by our gear and I swear I felt his relief just as if it was my own when he replied with an XO.

He'd been past a shop in his search for the horses, and as he returned at the jog he passes me a bottle of water just as the tears start flowing. They just seem to burst out of me as he wraps his arms around me and just stands there holding me. A shiver grips me as the scene replays again and again in my head.

I stare into Jac's eyes as he towers over me, torn from his back having lost the pack horse I lie defenseless in front of him, a 500kg horse whose eyes are wild and filled with panic.

I roll over in the hope of finding safety, looking up again only to see the back of Jac and Willow surrounded by dust as they gallop away. But the shock; what makes me sick to my stomach is not my own fall. No, what makes me shiver is the sight of Pete managing the stallion with a cool and calm head as it panics among the drama. He screams at me, making sure I'm OK and then pushes Bill into the Gallop to catch the escapees as I nod and sit up among our scattered belongings thrown from the pack horse and torn saddle bags.

As Bill takes that first launch to the gallop, my own scream cuts through the air...

"PEEETTEEEE! YOUR GIRTHS SNAPPED!"

That same second I see Bill hesitate, the saddle slipping in slow motion, and there's nothing I can do. In a split second Pete and his saddle ended up under, then off Bill who gently and carefully stepped over Pete, before bolting in the opposite direction. He could have been killed, should have been perhaps; but he is here, with me, holding me tightly assuring me everything will be OK. The tears subside and my rapid breathing slows, as both Pete, and waves of relief grip me. His hand moves slowly over my back, holding me softly, whispering to me what had saved him from serious injury.

Bill had warned him, hesitating the moment he gave him the cue to gallop, Bill didn't react immediately, as if asking if he was sure of what he was asking almost. That, combined with my scream meant Pete pulled him up, right as the supposedly top of the line girth leather snapped. Pete didn't care about any of the gear, he wasn't mad, but he was defiantly in pain, and again and again all he cared about was Jill and I being safe and unharmed. Pete had broken his hand, and I had some pretty impressive bruises caused by the lead rope, but overall, we were fine though still missing two horses and a lot of gear.

As the sinking sun started to color the evening sky in dazzling display of autumn colors matching the falling leaves drifting on the light breeze, Jac had been found by the same folks who brought back Bill... but Willow had disappeared.

We pitched camp only meters from the scene of the disaster by a tall stand of poplars and a small lake hemmed in on three sides by willows. I look into Pete's blue eyes... and we both know what fate awaits Willow, but we hope for a miracle. The reality is too unbearable... the knowledge that Willow will be in someone's freezer for food if not returned by tomorrows rising sun. The only way to deal with this right now is to cross our fingers and hope for the best. Pete and I have a sort of "silent conversation", an understanding, a moment where no words are needed.

We sit looking at the small lake that evening, laughing enjoying pasta with homemade tomato sauce, a leftover from the morning gifted to us by some friendly Kazakhs. We sit sipping cognac, another gift, and we both just sort of let go... because there is nothing else to do, and are both grateful for what we still have. Jill joins in too, coming for some cuddles outside, nudging me with her wet nose as if she trying to say, we're ok Mum.

But it all felt so familiar to me; this failure and picking yourself up straight after... and if I'm completely honest, this wasn't the first horse I'd lost to the steppe. That night as I drifted off to sleep, the dark memories of my Mongolian Horse trip came back to haunt me... forcing me to live through it once more, even after its pain had been washed away by all the beauty I'd experienced since. There I was some 3 months earlier; I was standing on the steppe with Jill. Proud as hell; holding MY HORSES saddled and ready to go! One last hug, a handshake, a last picture taken... and I turn away.

I'd been helped by a local not only buying the horses, but getting used to the Mongolian saddle and packing a pack-horse. Let me rephrase that... I watched him doing it. He hadn't even let me saddle my own horse I'd been riding with him for the past 3 days! I was too scared to ask; full of self-doubt and scared of failure. Not to mention Mongolian horses are as moody, aggressive and dangerous as the steppe they live on.

My toes slip into the left stirrup; my pony softly murmurs in anticipation... I swing into the saddle, and it begins. Afraid to look at my phone, I focus firmly on leading my pack horse and managing anxiety, fear and emotions.

I reminded myself it's not jumping in the river that drowns you... it's not swimming when you're in there. I'd just jumped in the river; and was so focused on staying afloat. Right of the bat I managed 30 minutes in the wrong direction, the first of many mistakes I'd make "learning to swim." I sigh... shame sweeping over me knowing I'll need to try and sneak back past my start point and hope I'm not seen and laughed at.

I rode a black gelding; pint sized "13.2hh", cheeky, lovable and his spirited walk seemed to almost go in tune to a song stuck in my head. The "Bear necessities", from the jungle book sung aloud on the empty steppe was where Mowgli found his name. He was a bit nervous and hard on the reins, clearly born to be wild and free.

My packhorse was a beautiful paint; calm and friendly... but bone lazy and always doing his own thing. Mulan... no; not a Rick and Morty Szechwan sauce reference. But the movie from the 90's; that's what I thought of. A Disney cartoon about a woman so brave and strong and by the end of the first day, I felt a little torn apart between Mowgli's forward nature, and Mulan's dawdling along behind.

I thought back on the seller laughing, shaking his head when he heard my plans. I looked at my horses, and promised myself to push through whatever may come and earn the respect of all these doubting Mongolians; but most importantly... myself. But it was far from the last time I was laughed at, or had people looking amused as I passed by. They would often come over when I was about to saddle up to "help me." I didn't know how to handle it really, maybe they were just trying to be nice?

"*Thank you very much, but I can do this myself!*"

Despite my false bravado, the first days I put the pack on Mulan were an absolute disaster, and I had to stop at least twice a day to fix or completely re-saddle him. I appreciated the patience and calmness shown to me by my boys which isn't something Mongolian horses are not necessarily known for. Was it my horsemanship? Or had I just got lucky with the horses?

Either way, Mowgli softened more each day as the scenery proceeded to simply take my breath away as Jill bounded happily along by my side. There's something inexplicable about the seemingly endless steppe; no fences, but hemmed in by mountains appearing as nothing more than a faint blue smudge in the distance. The hum of insects, soft murmur of creeks and the constant white noise of the wind moving the summer flowers under a cloudless blue sky filled me with a freedom I've never felt before. But it couldn't last... just as it didn't last in Kazakhstan.

A few nights in; barking dogs woke me up. I'd seen two stray puppies not far from my tent earlier in the day, so in my mind I put two and two together. But I was wrong. I heard restless horses; and still half asleep, I curse the dogs. But when Jill got up... I knew something was wrong. I fumble for a light, and slip on my shoes. I struggle half-dressed out of a cheap, one-person tent that's far from suitable into a world I felt I had no place in.

Once again I was swimming against the current in a river I felt could easily drown me. I swung my pathetic head torch about, and felt my entire being sink through my feet as I passed the light over the area in which I'd tied up Mowgli. All I could see was the loosely identifiable silhouette of two horses and a rider trotting away from my camp. I freeze... trembling, and as if I had to make sure; I walk towards the tent only to turn around to return to the spot my horse was just stolen from. I start to shake... panic... gasping for air. Thankfully Mulan was still there. He became my rock; I lean on him terrified the thief may come back?

As tears stream down my face, and I try not to think what could have, or still could happen; the night dragged on seemingly forever. I feel the only reason I kept it together was my friend Liz who stayed up all night on the phone to me. Liz, a girl I'd ridden with previously had planned to join me again; but this time hiking, and with her on the other end of the phone I kept my sanity. I nearly cried when she finally arrived the next day and we proceeded to find a hotel which allowed Mulan stay in their backyard; but what now?

I could quit... or I'd need a new horse.

Fortunately, the police officer who recorded Mowgli's theft helped us to find a new horse and set me and Liz on a life changing journey together. My new horse Luna was a stunning little palomino gelding, and it was nothing short of love at first sight for the two of us.

Four weeks later; I still hadn't learned much about long riding, horse gear, or outdoor equipment in general. But I had learned that all I had, wasn't made for a long-distance ride. In saying that, I had learned that even with the little we had, we could make it work; after all, the most important thing is the horses and the way your team grows together. The moments we shared made every low point, every tear, and every failure worth the risk; more importantly I constantly learned the lesson these things taught me. Cold weather, injuries or the wolves "which I still didn't know about yet", could have easily killed us.

Mulan had been led by Liz, and our connection wasn't that strong as a result. But Luna; he broke my heart. When I handed his lead rope to his old owner and an uncertain future, I literally felt sick, and in the days that followed I often broke down in tears. And now once more, this time in Kazakhstan, I have ridden the emotional roller coaster that is horsemanship, and once more I've had a massive setback.

As the first of the sun's rays hit the tent, I wake up somewhat startled by my uncomfortable dreams to find myself safe, by that small lake with Jill, our boys and Pete brewing me a coffee. The memory of yesterday comes back in one fell swoop, everything... I don't know what to feel, still as exhausted from the night just as much as from the events from yesterday. Pete turns to me with a smile and starts laughing and swearing... he might have ruined my coffee by accidentally buying sparkling water yesterday.

I feel new strength awaken, I've gotten up before and I will sure as hell get knocked down again. As I sit by our little lake watching the boys graze, Jill wade into the water and drink by biting the water like and excited puppy as the sun rises slowly through a cloudless sky; Pete tells me a motivational story, the first of many I'll hear when things have me feeling low.

"Lu, a mate of mine is in the SAS "Special Air Service"; they don't deliver mail, they deliver arse kickings to bad guys. Getting into the group is almost impossible, and the selection course is physically just short of impossible; but mentally it's even harder still. As a result, only the mentally and physically strongest soldiers make it. During the selection course many tell the instructors they can't go on... not one more step.

My mate told me, every single person that quits is told go put your pack and kit on the truck, you're going home; and right after saying they can't get up... they do. They do and walk to the truck just so they can quit. I didn't come here to use the last strength I have to quit, and I know you didn't either princess. Let's cinch up soon, it looks like we've got a cracker day to remind us why we came here and why todays not going to be the day we quit..."

District 12.

There have been very few times when we have had any issues with hostility towards us, and if I'm completely honest I would consider a night out at the pub in most parts of the developed world more dangerous than anything we have gotten up too. How often has there been a story on the nightly news of a "Coward punch" resulting in the death of a young man having a few beers after a week's work. In fact, I nearly lost my closest childhood friend to an incident like this. He was eating a pie, drunk at around 3am after a big night, when he was hit without warning and fell backwards down a flight of stone steps landing on the back of his head sending him into a coma for the best part of a week.

It's something people just don't think about really; just how dangerous life can be at home. America and it's gun violence, knife crime that's just out of control in the UK and an epidemic of both meth amphetamine and youth gang crime in my home city of Melbourne. But people view what we do as dangerous, and the truth is that's just not true. We need to be careful of course; in particular Luisa as she has got a pretty stunning figure and on more than one occasion has attracted the wrong sort of male attention.

But the biggest danger we face daily is of course our own horses. Horses are dangerous... that's putting it mildly, and I've seen first-hand several near fatal accidents involving horses. A young track rider lost it at the gallop while I was working in the stables only a few hundred meters away. He shattered his femur, and I do mean shattered. Bone was protruding and he tore is femoral artery and if it weren't for the outstanding and immediate actions of track staff, and the arrival of an air ambulance literally within minutes he most certainly would have lost his life.

In another incident the daughter of the folks that owned my local watering hole in New Zealand was almost lost during a ride. Her horse simply had a heart attack and dropped dead, rolling over and down a small bank crushing her under him as he fell. She jumped up in shock; mortified her horse had just dropped dead and wasn't even aware she had a punctured lung, and spinal damage that would see her in a specialist spinal unit for several months. But the worst; seeing our friend and this books editor Ali hit by a galloping stallion and trampled. To this day, I am still traumatized holding her limp lifeless body in the moments before we resuscitated her.

The point I'm trying to make here is this; both Luisa and I take care to avoid trouble, and when we see a chance there could be an issue, we turn around and quietly head in the other direction. You can't let fear control your life, and there is some sort of perception that people in the "Scary Stans" are hostile towards the west. It's simply not true, and aside from the odd drunk that got a bit annoying, we never felt afraid at all with one exception.

There was literally only one day on our entire ride where I was even remotely worried about the threat posed by other people; and... it's actually pretty hilarious. I'm not even sure how to tell it, more so if I should... but I will start by saying this. Do not in any way let this story put you off visiting these beautiful nations full to the brim with beautiful caring people. Instead, use it as a cautionary tale showing how a bit of humor, good manners, a bit of luck, and an ability to talk your way into or out of anything can mean getting out of a tight spot with a few laughs and a hell of a story to tell when you get home.

Luisa and I had been covering ground quite quickly across Northern Kazakhstan over the last few days, with literally nothing but an endless steppe before us dotted with small brackish lakes surrounded by a brilliant tapestry of blood red salt bush, Autumn flowers and the low grasses of the prairie. It was gorgeous, but we had a bit of an issue finding fresh water, so had decided to take a short detour into a coal mining town I have long since forgotten the name of. Google earth images hinted it should have all we need, as a decent size town sprawled beside a huge open cast coal mine I am sure you could easily see from space.

As usual, Lu and I ambled along chatting away as we approached what appeared to be the main part of the city, with Jill happily bounding about our feet. As we drew closer, we realized it wasn't in fact a city at all; but a huge sprawling cemetery made up of thousands of Muslim crypts, each around the size of a small building, with the city itself apparently on the other side of the mine.

The land changed from flat steppe, to rolling and steep square edge hills that had been formed by maybe a century of dumped mine over burden. It looked almost post-apocalyptic... brown patchy grass, rubbish strewn everywhere, with abandoned fridges, rusting over turned cars and other debris lying scattered among a sea of smashed vodka bottles and beer cans.

Luisa and I felt uneasy, and as a mare on heat with a yearling at her heels charged Billy and I; Luisa's time in the saddle mustering in Australia proved its worth as she challenged and bullied the mare, often charging it as I kept Bill hard on the bit with heels in his ribs, focused on the city ahead, not the mare. We pushed past without incident, but this was the first time I had ever managed a stallion near a mare in heat... looking back I laugh, as now it is simply second nature, but back then I felt like a sir surviving that incident without issue.

It could have gone wrong, as it had done less than two weeks earlier but it didn't. It didn't because bonds had been formed between us all, and in the months to come those bonds would change us all forever.

Luisa was carrying the bow today, and although far curvier and without hair she, looked every bit a real-life Katniss Everdeen as she charged and bullied that mare upon the little ginger gelding Jac. And this place; well, it couldn't have looked more like District 12 from the Hunger Games if it tried. It was utterly miserable in every way shape and form, in total contrast to Russia and places like Pavlodar to the North of us which were just so modern, clean, and in every way equal to the modern cities of western nations.

Here it felt like a post-apocalyptic failed soviet state in every sense as we skirted the mine in an attempt to find our way into the town. It was a Sunday; cloudless, warm and sunny as we found a nice sheltered spot among the overburden and rubbish to rest the horses while I ducked into town for supplies. I left Luisa alone with the horses and Jill, opting to head in quickly to check things out and return with the basics before finding a camp distant enough from this place to feel alone as we aimed to rest the horses the following day.

We were soon pushing off with the essentials, and Lu and I passed a can of beer between us as we departed into more movie set like wasteland. Broken concrete lay in endless piles pushed by bulldozers decades earlier, and more garbage lay scattered, thrown callously at the towns edge at seemingly every place a vehicle could be driven too. Lu and I had a bit of a laugh as Bill the Bastard got his foot tangled in a long tinsel Christmas decoration and just continued walking as though nothing was wrong. I leaned off him, hanging from the horn with my left hand unhooking it as he walked, springing back into the saddle to greet the beer being handed back to me by Luisa.

It was the little moments like this where Bill and I were really starting to bond, and every remaining aspect of a western mindset were leaving me. Luisa was growing as a horseman too, but her head was still getting in the way as she worked with Jac; like a good stock-man would, not a drunk fool on a carefree adventure. We pitched camp in a fold of land by a tiny pond with good grass, and a few small willows a good 6 kilometers from town. It was perfect for a 36 hour stay, and we were hidden not just from view, but from the wind and light rain that greeted us the following morning.

We needed cash, we needed a new iPhone charging cable "I've been through 13 in the last year" and to charge up our power banks and my iPhone. With the horses free grazing but hobbled by good water, I left Luisa in bed with a cup of hot chocolate, writing in her diary shortly after sunrise to head in to town alone once more. It was a long way, maybe a good hour or more through mist and drizzle across dried mud flats covered in more blood red salt bush with caked salt watermarks in long lines running among its stems, broken only by the cracks in the mud. Today was cold, miserable, and my kind of day for a long walk with ear buds in cranking a bit of "Hill Top Hoods".

I was soon in town, and found a phone shop easily enough... but everywhere I went I felt uneasy eyes on me. Apartment buildings, all were in varying states of disrepair, crumbling, painted in soft weathered pastel tones lined the dirt streets with cows wandering too and fro, up and down. Trees separated the street, if you could call it that, from the cracked, patchy concrete foot paths; and all the gnarled, sickly looking trees sported a band of white paint from their base to about shoulder height.

I found an ATM, secured more cash and messaged Luisa as the battery on my iPhone warned less than 10% remained, telling her I'd be on my way back shortly... but that was not to be. As I attempted to acquire groceries, a seemingly friendly local began demanding I return with him for a drink, and that I may charge my phone and power bank at his property. Things felt wrong, but to refuse felt as though it would carry more negative consequences, so I reluctantly complied.

It didn't take long for me to realize that I had been nothing short of kidnapped by the former Russian mafia as we chatted in the basement of an electrical supply store, littered with clutter and recycled junk in various states of decay. They all wore a uniform of sort; ADIDAS tracksuits with trainers, or dark jeans with leather jackets and sunglasses. I was forced to drink vodka; accused again and again of being an American as did my best impersonations of a kangaroo, and heard story after story of their time escaping Russia after the fall of the Soviet Union.

People came and went, and I got drunker by the minute trying to find the perfect opportunity to get out politely, when a large amount of weapons began to appear including a US issued M249 SAW, a belt fed section automatic weapon, most likely captured in the Afghanistan conflict. Ahh... righto. Things seem to be escalating fast... but this was fortunately to my advantage. You see I know the SAW very intimately, having carried the New Zealand equivalent on deployment and showed my new friends how to field strip and assemble it with speed and efficiency far beyond their expectations while denouncing the US as a hateful oppressor of many nations.

Vodka, and I do mean a lot of vodka, followed while my phone charged and I planned my escape. Although somewhat terrifying, I had the confidence of a bunch of vodka on my side, and these guys had now well established I was worth no value as either a hostage, or for that matter had any current or previous beef with them at all. I was just a curious oddity from a land that had kangaroos and no interest in soccer; but the tension in the air could still have been cut with a knife as I slipped my charger, phone and power bank into my pocket and suggested we go out for a cigarette and some air. They joined me, and we joked "uneasily I might add" as I looked for the perfect moment to cut and run with my small day bag of supplies.

We were in a small courtyard among the apartment buildings, a small sad looking sea saw and swing set rusting among other depressing reminders of better days, as drizzle fell from a grey sky, and Autumn leaves swirled about in the Autumn mist. And then I saw my chance; the others ducked back in to grab more vodka... and I was gone. I was piss drunk, but cut a trail as fast as I could across town and towards the dump Lu and I had ridden through the evening before.

Looking back, I saw a 90's era black BMW approaching so quickened my pace, ducking into another small general store for cover; unaware I had jumped out of the frying pan and into the fire. I collected a few items, another vodka and hastily messaged Luisa saying little more than I am hammed drunk, the town is full of mafia and I'm pretty keen to get out of here. This is where it gets really weird... and I do mean really weird. The shop keeper spoke perfect English and as he rung up my items, he chatted away...

"Don't go back outside, mafia is waiting for you, come back to my place and smoke some weed instead, it's the best, I bring it back from Afghanistan from when I fight Americans..."

So... This is awkward. Armed mafia on one side of the street wanting me to drink vodka with them, and now the Taliban want me to get high with them.

All I knew was I had to protect Luisa. They knew we had ridden here, but they had no idea where we were camped, and now I needed to ensure I could make a clean break in such a way I was unlikely to be followed by people I may or may not have pissed off. To be completely fair, they were all probably just doing their very best to be friendly, with intent lost in a haze of vodka and cultural misunderstandings.

So... a few minutes later I was smoking bucket bongs with a former Taliban fighter in his "man cave" with weed chopped up on a playboy centerfold featuring Madonna from the mid 80's. He wasn't a bad dude to be fair, keen to show me piles of vintage porn and talk about how his three wives all hated each other, only making the sex better as he would sleep with whoever was angriest at him. By now though, I am now utterly off my face; both blind drunk and very, very, high.

To be honest... during my walk into town I saw things going differently.

I was then invited to lunch; and obliged as to be honest... it's not like I really had a choice while trying as best I could to maintain any sort of coherent thought. I clung dearly to my bag of supplies, and did my best to message Luisa something that made any sort of sense without much luck.

But at some point, his wives started going at it, and I do mean fully going for it! There was yelling, slapping and all sorts of carry on as I sat there utterly on another plane, looking and feeling like something out of Fear and Loathing in Las Vegas. I look across the lunch table, a generous and delicious spread of local foods, and see the Taliban fella laughing as the ladies erupt into yet another scuffle and for me that was it.

I'm barely able to make the door, but I do, slinging my bag over my shoulder, mumbling unintelligible rubbish as I burst into the unfamiliar city totally devoid of landmarks I recognize. Drizzle obscured visibility... oh, and I am utterly wasted; so, I just aim for where I assume the edge of town is and I leg it.

By some miracle, I get clear of the city; rapidly disappearing into the mist and soon finding myself alone and out of sight of any features on that same bleak, desolate salt marsh flat I'd crossed on my way into town? Google earth... that's what I'll use. I'll find Lu using google earth. But I can hear vehicles coming... or maybe I can't? Maybe I'm so high I am paranoid? So I run, diving into a small ditch like a ninja.

Next, I scan the horizon for threats making pretend binoculars with my hands and some sort of "beep beep beep" radar scanning noise as I think back to my break contact drills for evading enemy whilst training to be a reconnaissance trooper in the army. Moments of clarity drift by me, and in these I manage to figure out vaguely where Luisa is... and I jog, run, in a shambolic stumble towards her cutting left and right to confuse my imagined pursers for what seemed like hours.

It was NOT my finest moment by any stretch of the imagination, and once safely in the fetal position beside Luisa and Jill... I swore that was the last time I would ever drink with the mafia or smoke bongs with the Taliban in order to escape a post-apocalyptic soviet era coal mining town...

In all honesty; this story is such a blur, and although I drink a bit of booze, it's very rare for me to hit it hard enough to make a mess of myself, and very, very, rare for me to ever use drugs. Maybe this story is born of drug and alcohol obscured paranoia and exhaustion... or maybe it happens just like that, I'll never know. All I know is I'd much rather keep my wits about me, and this was the only time I couldn't avoid drinking due to peer group pressure in a strange land.

Beersheba.

I name all my horses after either famous battles, or famous Australian war horses with Bill the Bastard named after an Aussie legend of such stature... he has his own statue to commemorate his courageous and also caring efforts on the battle field.

My Bill, well, he was named perfectly. He was a Bastard, but he was my Bastard and he would have done anything for me and with me. If it were humanly possible, despite his sometimes clumsy nature, I would have given anything to keep riding him all the way to Europe. But crossing the boarders with horses is simply not practical, affordable and reasonable to even attempt on our time frame and budget. So, sadly, we had to sell him and Jac before crossing into Uzbekistan. After two months with Bill, I was so heart-broken I couldn't even bring myself to properly say good bye to him... and I thought back to the question I had been asked again and again by people before I left for this trip.

"Pete, you're going to do something pretty extreme with haphazard preparation at best, and let's be fair... you haven't exactly been born in the saddle. So, what's your biggest fear?"

My answer was always the same. Saying goodbye to a horse I have fallen in love with. I knew it was going to happen, and it has happened several times to me since I left Australia; but I was prepared for it and there's sadly just no way to get around it. Luisa suffered similarly with her little Mongol horse, and I'm trying to not take too big a "spiritual" path with my ramblings... but there is most defiantly a spiritual connection of some unseen energy that can pass between a horse and its rider as energy passes between lovers.

Think about the touch of a lover's hand on your shoulder after a long day, think about looking into the eyes of your Mum as she unwraps a thoughtful gift on Christmas morning that has tears well in her eyes before she throws her arms around you, and thanks you. These feelings, this shared energy is absolutely real between people and horses if they are open to its existence and choose to allow these feelings to flow between one another. And then one day, you have to just cut that thread, turn, and walk away.

Digger remembered me the second he saw me the first time I revisited him a year after we met, and I have no doubt Bill would run to me the second I saw him, even if he were running wild on the steppe with mares. Yeah, it hurts. But as they say it is better to have loved and lost than never to have loved at all, and being able to train, use, and then switch horses means that you can cover country that is otherwise out of reach to you due to boarders and physical limitations of a horse or lameness.

Lameness, another of my biggest fears; that along with high speed traffic and long bridges. These were my other biggest fears and I have come to terms with all of these and if there's ever a time to write cautionary tales about these, it's here. I have come to terms with these risks far better than I ever though I would have to be honest, having dealt with a few near misses and some pretty tense moments in which I both grew as a horseman and also grew in experience of what needs to just be avoided all together. But still, as any horse owner who's parted with a beloved partner will tell you... there is nothing that can prepare you for your separation from a horse.

We will be eternally grateful to the nation of Kazakhstan for our time there and its treatment of Luisa, Jill and I, and are happy knowing we left our Kazakh horses with Kazakh horsemen to see out there days doing what they love most and were born for.

After such an amazing time there crossing into Uzbekistan held a little uncertainty, but we were welcomed with open arms just as we had been in Kazakhstan and through an add Luisa had placed on couchsurfing.com we were able to meet with an English-speaking native that provided us with endless help and support. More about him later, this is a story about failure, not the kindness and warmth of strangers.

We needed two new horses, and after checking out several we made what ended up being two poor decisions. Luisa's horse Bluey was a stunning, tall, powerful Karabair stallion that had just entered his retirement after years as a Kupkari sports horse. Sadly, he carried a few niggling injuries that were to plague our trip with veterinary stops. But for me, in part due to the language barrier, and miscommunication with Luisa, we chose a completely wild, unbroken young stallion.

I came here to test myself; I didn't come here to get a trekking horse and see the sights, I came here for a wild and crazy adventure with twists and turns that would see me tested to my very core. And in Beersheba, I got that and more. But it was a poor decision and that rests firmly at my feet and what I did was both foolish and beyond the capabilities of the situation we were in. It wasn't so much that I couldn't manage Beersheba, on the contrary I managed him and tamed him exceptionally well given the situation, but he was wild, terrified of people and all in all a very dangerous horse.

I keep saying given the situation... why? Because if I had the time to spend a month with this horse working with him daily in a round pen, I'd have had one hell of a good horse, and if I'd known him since he was a foal I'd have had a tame pet like Bill with all the power and grace of a Karabair sports horse as well. But that wasn't the situation at all; I had two days to let him settle quietly in a stall after he had been tethered in a paddock for years before. But the Uzbek's didn't see a problem with his suitability, so I decided to put myself to a test that could with hindsight have easily killed me.

That first ride... hell, it took three people to hold him down enough for me to even get on him! I treated him like a Mongol horse, turning him in tight circles as Lu mounted Bluey and then, that was it, we were off. Straight into a snow storm in the middle of a city. Here's where it should go wrong right? Well, it didn't. I managed that little stallion like a champ, and even crossed a four-lane motorway at the canter within an hour. Believe it or not, his mouth was soft and he required very little input from me at all as he chose to follow the taller more powerful lead stallion Luisa rode.

But he was dangerous, he was so full of fear and it took all my skill to stop him spooking on numerous occasions, a habit of his that would nearly get me killed a few times. It's fair to say that despite being of a beautiful physic, Beersheba had been abused, I'd say in part due to his temper and unwillingness to let people come anywhere near him. But, in those first few days I got him riding beautifully. To ride, he was just breathtaking with every gait smother than the last, and his gallop was just dream like.

To his credit, he only really became dangerous when I tried to work with him on the ground. Putting his saddle on and off was a delicate and dangerous affair much as it is with a Mongol horse, and although used to this kind of horse and this routine, not to mention supported by the amazing, talented, beautiful and courageous woman by my side... there is no place on a long ride for a horse of Beersheba's temperament.

He was purchased because I wanted to ride a powerful pocket battleship for egotistical reasons and that was in every way a ridiculous decision to make.

Luisa, to be fair, backed me in it; but ultimately it was that primitive part of my male brain that locks eyes with a dangerous wild creature and thinks... I will tame this beast that put us both in danger. I was always sort of "managing" Beersheba when we rode, never able to relax and just enjoy the scenery as I had with Bill or Jac, and there was no way Luisa could ride him. Jill and Beersheba were a dangerous combination at the best of times and Jill rarely left Luisa's side while riding.

He would do things, especially on roads where he would just freak out over the stupid things and try to bolt or turn into traffic. Jill didn't help, in fact Jill was a bloody nightmare. It was by far safest if Luisa held Beersheba when I got on and off, as the slightest movement or noise he didn't expect could send him into a panic. This is how he hurt me for the first time, I'd become complacent. I am used to horses like this and I am so used to the risks that I didn't give things the correct respect they deserved, and as I slipped my feet out of the stirrups, Jill startled Beersheba. He spooked just a little, but Jill gets excited when this happens "worst cattle dog ever", and decided to help by barking and pushing the horse I was half off with limited control of. He bolted; I bailed, and dam did I hit the ground hard.

My fault. My fault for so many reasons it's not funny. I needed to prepare the horse better, I could have asked Luisa to hold the horse as I got off, I could have asked her to control Jill better, I could have moved away from Jill and Luisa and I could have purchased a more suitable horse from the start rather than being such and egotistical cowboy about the whole thing.

BUT... I was learning lessons few people are ever game enough to learn in this way. I was in the full speed ahead crash course of wild horsemanship doing university papers with a kindergarten education behind me. All in all, I was doing very well, but still this was a foolish decision for me to have made in taking this gorgeous horse. And he was gorgeous, I rode him up and down stairs, across foot bridges, train tracks, through a huge concrete culvert, across motorways and even literally into the heart of a city. He was becoming a tame pet, and had a beautiful side to him... but that fearful wild side still persisted, meaning disaster always only a split second of poor judgment on my part away.

There were moments when it was only Luisa's exceptional horsemanship that got me through tough spots, and as much as I have helped her at times, it was with Beersheba where she shined brighter than I think she will ever realize.

Uzbekistan was a stunning place to ride, but we were plagued with lameness issue with Bluey, and although Beersheba had settled to a superb degree, any small setbacks with Beersheba were really held onto by him. We had him almost where he needed to be for safe enjoyable riding when we made the foolish decision to have the horses shoed. Well; Beersheba's trust in me, people and everything else went right out the window again, and he wasn't the same horse in the days that followed. Sure, he was great on the steppe and would bravely charge at the Shepard dogs trying to eat Jill, but would spook at Jill...

And then it all fell apart. Beersheba slipped on flat grass and fell in slow motion as though sitting down, pinning my right foot in the stirrup as I gently went to step off him. I don't know if Jill got excited or not, I can't remember and I am not going to try and pass the blame to her here either, but the long and the short of it was I ended up being dragged at the gallop with poor sweet Luisa thinking she had finally seen me get killed by this bloody horse she had come to both distrust and dislike.

Same as before, the only thing really hurt was my pride and after a bit of ground work and of course getting back on, "same as with the other incident" we settled in for the night. We didn't know it then, but it would be our last night on the Uzbek steppe.

A beautiful clear day dawned, we saddled up, and began walking the horses; a tradition we do each morning to get the horses and us warm before beginning the days ride. We were soon following along the edge of a canal running parallel to a town separated from us by cultivated farmland and as usual we had our girths looser than we would when riding.

As we passed by a shepherd tending his goats, Beersheba's saddle slipped to the side a little which sent him into a panic... I tried to control him as best I could, but he began bucking and kicking wildly and of course Jill decided to help making the whole situation turn from bad, to utter catastrophe. As I was being dragged across the wet grass trying to find my footing I slipped forwards, and in the split second before a kick from his wildly flailing hind legs struck my face; I put up my arm just enough to save what could easily have been a fatal hit to the face. Luisa was screaming at Jill and managing Bluey, who for the first time since we had him had too lost his composure, most likely still traumatized by the terrible job the farriers did in attempting and failing to shoe our horses.

Like lightning Luisa had cinched up, and was after Beersheba as once again I lay in pain looking at the utter scene of carnage before me. Gear was everywhere, and Beersheba was still madly pig rooting and bucking a good 1000 meters away. I was punch drunk, but OK and a glare from me at Jill let her know in no uncertain terms she wasn't far off being thrown in the canal.

I figured I'd pick up the gear on the way back, opting to go get my horse first that thankfully Luisa had now hemmed in with Bluey. He didn't want to be caught; not even by me, his best mate who gave him hugs, carrots and scratches, choosing to just glare at me and warn me not to come any closer. Minutes passed; and he relented, having calmed down enough to realize he'd made a mistake and allowed me to pull the quick release knot dropping the saddle out from underneath him as I took control of his lead rope. This was my fault. All of it. Not his.

We should have taken the time to go through slipping the saddle underneath him as I have done since with every single horse I have trained from then on. I could have done better, and I am doing better as a result of these lessons hard learned in the deep end of the pool having jumped in feet first fully clothed and a bit drunk. Bluey had slipped galloping after Beersheba too, and was limping a bit. That and I was very short and rude to Luisa due to the pain I was in too, for that I'm sorry, and I'm sorry that even despite of the 99.999% awesome time I in particular had with Beersheba, he simply was not the horse for this journey.

We walked the rest of the day, calming and reassuring the horses before being invited to stay with some amazing Uzbek's who came to care for us and the horses for some time to follow due to both Bluey's lameness, and the delayed concussion that followed for me the following day.

But there were still hits to come in Uzbekistan. Jill was attacked while walking alone with me to town two days later, and as I dragged each huge 30kg + shepherd dog off the terrified, screaming and bloodied little 17kg cattle dog, another would take its place savaging her. I was left with no choice other than to literally use an attacking stray as a club against the others to save Jill in what equated to nothing short of the most terrifying and rage fueled hand to hand combat of my life. By some miracle I turned the tide that was three dogs against one after throwing one over a fence, another into a fence and the other into the ground, scooping Jill up into my arms as she screamed and cried in agony.

I took her home, and for the next few days treated her wounds and tried to figure out how on earth I had managed to get out of the situation without so much as single scratch on me. It had really rattled us, and with three of the five of us out of action, things needed to take a definite pause so we could re-group before our next move. I will always have a very special place in my heart for Beersheba, as bonding with him taught me more than a quiet horse ever could.

The grey ghosts.

If you are easily upset by stories that involve the death of horses, and at times honest and graphic details of incidents involving the unpreventable loss of horses to both wolves and illness... please skip the rest of this chapter.

This story starts back in Mongolia; with the first time we noticed the grey shadows of the steppe. Wolves. Wolves are one of the world's most misunderstood creatures. For thousands of years they were demonized and feared, and if you ask me it was for very good reason.

They fear man; but they are much like a domestic dog in that when they feel they can get away with something, they will. A dog will sit and stay by a bowl of food without eating it so long as the owner is watching; but if you turn off the light and the dog feels you can no longer see it, the food will be gone in seconds.

Wolves stay out of sight, but they are present far more often than I like during our wanderings on the vast Asian steppe. And just like a dog staring at a food bowl with their owner present, the wolves are just waiting for the light to go out. Sadly; I know a man who lost his grand-children to a wolf that ran them down in deep Mongolian snow as they tended to their stock. My Mongolian Horse, "Digger," killed a wolf that was attacking him by biting and trampling it to death; the same wolf had killed 5 of his herd that week.

All too often we hear them howling in the darkness; occasionally glimpsing a flash of grey fur in the brush or the glint of their eyes in the glow of the campfire. But for the most part; they are the ghosts in the darkness that keep us huddled close to the edge of civilization and the perceived safety it provides. We respect them; admire them, and know them in their pure, wild and deadly true form.

But "National Geographic" tells a different story about wolves. They tell a story that sells the image of a "spirit animal". An image typically found on the tie-dyed T shirt of an over-weight woman shopping in K-mart at 2pm in pajama bottoms and slippers. It's an image you'll seen the world over in shops selling incense, hackie sacks and gypsies treasures; the majestic wolf howling at the moon with native American images and dream catchers surrounding it on a poster.

There is a "mystique" surrounding them, and the enduring myth that they are very rare and special which has seen them reintroduced and protected in much of the developed world.

Personally, I think this is great; I love that wild creatures flourish in wild places and are afforded the protection they deserve, but there is a side to this conservation success story that remains unseen. Wolf numbers can rise very quickly, and the rate at which they deplete food resources is astronomical; and it's when this occurs that they come into conflict with man as they have done for thousands of years. Balance is essential in nature; and second only to man, the wolf can quickly and devastatingly effect delicate balances in an environment. I seek not to demonize them, I just aim to present the cold hard facts regarding those we share the forest and steppe with along our journey to Germany, then Australia... together.

When we arrived in Caucasus, it was the peak of winter, and although much warmer that the steppe of the Stans, it was still frozen and devoid of color. A greyscale land of rolling mountain foothills covered in drifting snow, forested valleys masked by eerie mist with the muffled clang of cow bells and muttering stock rising upwards. It had a sort of "1920's" black and white film feel. We loved it here; and we jumped at the opportunity to settle into a soviet era stables and fix the place up while riding out the worst of winter.

The stables were located just below a tiny village of eight families perched of edge of a sprawling wilderness adjoining the mountains; the geographical boundary that separates Europe from Asia. It's a wild place, full of wild people that had as much in common with the modern world as they did with the old and the wilderness that surrounds them. Our nearest neighbor, "Mari" was the perfect example: twenty years old, a gorgeous leggy brunette taking Instagram selfies in designer clothing... while tending the family's cattle in the wind-swept fields under the ever present gaze of unseen predators.

I drop to a knee and close my eyes as I remove my medical gloves; fighting back images that still haunt me from the battle fields now long silent. I fold the gloves inside each other and drop them among the other medical rubbish littering the blood-stained cement floor of the soviet era stables.

There is no smell of cordite, no spent cartridges, and my ears aren't ringing from an exchange of gun fire that left screaming wounded in its wake; but the empty feeling inside... the welling sadness... it's exactly the same. Instead of the sound of vehicles, team mates and distant gunfire; there is just the soft murmuring of a mare who leans towards me, gently nudging me causing me to lose my balance. Lu stands beside me, and quickly steadies me with a hand on my shoulder, softly whispering...

"it will be OK Pete, you've done all you can."

Her words mirrored the mares gentle reassuring nudge seconds earlier, as we tried desperately to treat her foal's injuries as a result of last night's wolf attack.

Empathy. It has its down sides. I feel the mother's pain in every single part of my being; and worse still, I feel that dark sleeping bear inside me stir. Sure... this is nature. But nature for most is the discovery channel and right now I need to go get a bucket of warm water to try and get all of the blood off the floor and clean up the wound dressings, make a run to town to try and get medical supplies in boxes marked only in Russian while a veterinarian on the other side of the world messages me over Facebook advice on how to treat the wounds the foal was lucky to have survived.

Cowboy up yeah... sometimes it's not about getting back on a horse at all. It's about caring and doing all you can to save a life even though you've got literally no idea how to treat a horse that's just been torn apart by a wolf. You have no choice, you find a way, and when it's all settled down and you've done all you can; maybe then you think about going and picking up a rifle.

I named the foal "Straya", as he was a little battler until the end. But bathed in warm spring sunshine three days later, with Jill sitting by his side and her mother grazing in sad but knowing acceptance nearby, Straya slipped away peacefully. In the coming weeks another two foals were badly wounded, and a mare was savaged; with the wolves attempting to rip her unborn foal from her. After many hours of difficult labour, it was clear she had lost the foal which was now trapped inside her in her weakened state.

An older mare, she had once been a champion race horse; now, she lay on her side bleeding from her wounds under her tail. Once more we called a veterinarian on the other side of the world and asked for guidance, and once more my skills as a well-trained medic were again being used in a desperate attempt to save a horse.

I had no idea what I was doing, and the idea of reaching inside a horse to help pull a foal free was something I'd never imagined doing, particularly while being directed by a veterinarian in Argentina. The foal had not turned, and despite my best efforts at turning the foal so it could exit head first, I was unsuccessful. Instead, the foal was coming out hind legs first. I'd had my arms literally up to my armpits in an attempt to turn it, and was now instead dragging it free with all the strength I had. The tiny premature stallion slipped free finally, and the mare's relief was evident as her eyes met mine, sharing the sad understanding of her loss together.

We sat with the mare for some time, feeding her handfuls of hay, offering her water and gently stroking her neck as she sat calmly. I took the foal to a secluded valley just before dusk; laid it gently under a stand of poplar trees. The sleeping bear inside me had woken, and I placed two heavy 00 buckshot shells into the chambers of a shotgun; found a quiet spot to sit, watch, and wait for the predators that had ravaged the herd...

Maybe it was just a dream...

The wind howled outside, the voices of the owners disappearing with each step I take deeper into the dark, crumbling stables, lined with fine spiderwebs and stray nails hammered randomly into every beam as a hook, most never to have been used. Those old soviet stables in Georgia had me feeling like a child peeking into the lounge on Christmas morning, all full of hope and excitement. Most of the stalls were empty, the rest were piled high with bales of hay and cluttered junk collected years earlier, now just lay gathering dust.

The horses ran free in the hills surrounding our little valley home, dusted in fresh winter snows and fallen leaves, but there were a few in the stables out of the cold. Among the beams of light filtering into the dimly lit stone building through the countless holes in the roof, a curious chestnut face appears, stopping me from turning back to follow the owners and Pete outside. I make my may way towards him, slowly, calmly mumbling. With each step I drew closer, he backs up, so I stop just in front of his stall. Neither of us move; and the questions written all over his face tell me more than enough to know he is a young stallion.

He moves his head from one side to the other, then nods and murmurs before finally deciding to take a chance. As if in slow motion, he makes his way towards me, stretching his neck forward to sniff my hand. Little beams of light hit his coat, lighting it up in brilliant dappled golden hues. I smile, looking at the hopeless effort made to patch up the roof with a bit of plastic, half ripped off in the wind, now dancing in the drafty stables as it clung grimly to the rafters.

His figure appeared delicate, but strong from years growing up exposed to harsh environments, wolves and the endless search for new pasture in the hills that are near barren much of the year. I was daydreaming almost, and didn't even realize when he took that last step forward. He was now standing right in front of me, lowering his head, allowing me to touch his forehead; a glimpse of the big heart inside this stunning horse standing in front of me. As we touch, something inside me changes forever...

A voice calling my name breaks through to me, and we both take a step back; the moment is over. He turns back to look out of the narrow window in the crumbling grey wall, starring into the snow and windswept empty space, clearly longing to run free. I turned around, still smiling made my way back to the others. From here, the deal was we will break 3 horses for a small trekking business, fix up the stables a bit and take care of the horses during the worst of winter; and in exchange we can stay for free, and ride around Georgia.

All I cared about was if the Stallion was one of the breakers, he was going to be mine, and the moment that was confirmed, I knew we were in the right place.

Pete took on a violently aggressive little filly he named "Pippa", and he encouraged me to work with first a gorgeous fluffy little colt called "Kashlama" before taking on the stunning chestnut stallion named Alarmi. This... well this is the last thing I had dreamed would have been how I rode out winter, and for me it was literally a childhood dream that was about to come true.

We would have the stables to ourselves, but before the owners left we got the other horses in. I could barely hold back how excited I was as we mustered in the horses! It was magic; to muster cattle is one thing, but a whole heard of horses! Galloping among these majestic animals, their flying manes, their playful pig rooting, bucking, and the sound of thundering hooves had me almost in tears of joy. I could already see myself flying over these same fields with Alarmi, despite not even knowing where to begin with the training!

Alarmi called loudly from inside the stables as we approached, attempting to get the other two youngsters in. The Colt wondered in without so much as a complaint, but that dammed filly took hours. She was wilder than anything I'd seen yarded before, and I was more than relived to know she was Pete's problem, not mine. So, with the breakers in the stables; they left us to it, and as their four-wheel drive headed up the rough clay track towards the city, we did the first feed and water round.

That evening as we warmed our hands over the little oven beside our small bed, both wrapped in our sleeping bags to fight off the cold, I could still barely contain my excitement at the reality that mere meters from me were a pair of wild horses I'd be breaking to saddle!

Without Pete's help, constant motivational talks and his belief in me; I would still be standing in front of those stalls dreaming. Hell, without Pete saying "Hi" to me and Jill, I'd have probably just headed home to Germany and on too Australia via other means.

But now here I am; getting to know my boys, touching them and building trust, and it was all just happening naturally after that first step had been taken. I really can't stress enough how just deciding I was going to do this, fully committing, and pouring my heart and soul into these boys just made every step we took together so enjoyable. I honestly feel they felt the same, and under Pete's watchful eye, I put the skills I'd been picking up since childhood to work and right from the start it all just seemed to click, with our 5% improvement goal being achieved each and every day.

To get a head collar on Alarmi was no hassle at all, curious and trusting he was soon able to be easily led to the nearby waterhole and from pen to pen. Kashlama, "Kashi" on the other hand was only two; a full year younger than Alarmi, and still behaved like a foal around the other horses and was at first reserved around me despite being a friendly horse. I walked quietly, confidently and calmly into his stall, and the dance started. His big googly eyes were unsure, soft, kind, and almost reminded me of Jill when we first met.

As the hours, then days went by he stepped forward slowly but surely when I approached and would calm down very quickly, so much so I never had to force a single interaction. In total contrast, Pete had roped is filly day three and made it very clear he was in charge, not her.

This was the last time she fought Pete; he'd sat talking softly with her long enough, and now for the first time she was head collared and calming down. Pete just laughed and started throwing saddle blankets over her, rubbing her neck and brushing her as the fire in her eyes began to slowly soften towards the curiosity shown to me by my boys.

I run my hand through Kashi's fluffy brown coat while he drinks from the small frozen pool above the stables having just smashed the ice for him. I love it here, and I daydream, staring into space as he drinks and paws the freezing, muddy waterhole. Kashlama was great, lacking Alarmi's power, he was a gentle sweet little horse and it was almost as though he gave me just as much comfort and security as I gave him. His lead rope would hang loosely as he followed me around the stables, often asking with a soft nudge of his muzzle for pats and scratches before letting him return to his stall.

Everything with Kashi was so simple, but excitement would grow stepping into Alarmi's stall, and yet he hardly moves; instead choosing to call loudly to his mares while staring out the window. He doesn't really enjoy the long strokes of the brush along his back and down his flanks, but he stands still; accepting it and almost hinting its maybe not so bad. Soon I had both my boys used to being saddled for a few hours each day, and both were easily accepting the bit and bridle as was Pete's crazy little filly who was now almost as loving and soft as Kashla; but Alarmi was still stoic and self-assured tough co-operating none the less.

Ground work took us a while to find our rhythm and move as one as Alarmi didn't understand what I wanted from him, and would starts cantering and mucking up before even considering trying to understand what I was asking of him. But still, I felt some unspoken magic between us whenever we worked together, after all, he was such a teenager! Full of testosterone, a clown, and at times I could feel frustration creeping in; then, as if on cue he just steps towards me asking for love. He was cheeky and innocent, and there was little more I could do other than just laugh and send him back to work.

I was merciless and consistent, and we learned to speak a common language both through ground work and our long walks to the river each morning. It brought us closer together, with the lead rope leaving him enough space to be a child, but a child with very firm boundaries and as a result our love grew daily. Eventually, the time came to ride the boys, and as butterflies did loops in my stomach, I did up the chin strap on my helmet and stepped into a bright crisp winter's morning.

We had no round pen, just a flat area of frozen dirt dusted with light snow to the rear of the stables where Pete stood with Kashlama, saddled and waiting. He was calm after a morning of ground work and pats, and allowed me to gently place my whole body over his saddle as he had done many times before. As I slipped off, rewarding him with a gentle rub of his face as Pete stood quietly holding his lead rope. All those childish dreams, all those years learning and here I was, slipping my foot into the stirrup of a wild horse's saddle for the first time.

Pete nodded, smiled and I swung my leg up and over, finding the other stirrup effortlessly. And... nothing happened. Kashlama stood quietly waiting for me to give him his next suggestion as a rewarded him with pats and soft reassuring words. Pete smiled, saying nothing as he stepped back allowing the lead rope to lengthen to a distance of some meters. A soft kick, a click of the tongue, and just like that Kashi started to walk in circle for me as though he'd done this a thousand times before!

I could barely believe it, and I didn't want that first ride to ever stop. In less than a week, there was no need for a helmet, no need for worry of any kind as Kashi could be ridden softly, effortlessly and as if he'd been my partner for years! I was soon trotting, cantering and just loving the little guy as he was just so easy to work with; but still, in my heart and in my dreams it was Alarmi I longed for. I knew full well that Kashi was the perfect stepping stone towards that stunning stallion, and our first ride that was mere days away. I grin every time I think about the first time I rode Alarmi; those crazy first few seconds in which all I remember was Pete's voice yelling

"STAY ON!"

He'd given no warning at all, and had calmly let me on just as Kashi had, but... he decided his first step with me in the saddle would be towards the sky! DON'T LET GO! That was the only thing in my mind, I didn't, and in a few seconds Alarmi figured out Pete wasn't letting go, and I wasn't coming off and he just suddenly stopped. I hadn't fully grasped what had just happened, but I had sure as hell felt his power and in that first test of wills, I had won.

I hadn't lost my nerve, but after a short while sitting in the saddle as Pete and I calmed Alarmi, he became settled, and that same cheeky boy I'd fallen so deeply in love with. I slipped from the saddle without incident, and Pete, somewhat annoyed that Alarmi had made him spill a perfectly good can of beer, wandered back inside for a fresh one. Pete soon returned, and I was already hard at work with Alarmi trotting messy confused circles around me on the rope.

I smiled, Pete as usual had the portable USB speaker cranking Aussie Hip Hop as he sat watching on silently drinking a beer and giving me the confidence to keep at it and see this through. That smile grew from my normal smile, to my real smile... the crooked one reserved for special occasions as if as though on some sort of divine cue, Urthboy's song "The Arrow" which had moved me so deeply on the steppe of Kazakhstan begun playing. Those lyrics,

"I'd never had a better fu#$%^g time in my life."

They rung true watching this stallion circle me, slow to a stop, walk to me and bow his head for a pat. I turn to Pete, still sitting quietly with his beer and say...

"We're ready Pete, it's time to get back on."

Alarmi never bucked, never pig rooted or so much as disagreed with me ever again.

If I close my eyes, I can see his face just in front of mine, asking for treats, pats, affection and to connect with me. I can smell him, feel him as I lead him home from those high pastures above the stables as "my partner" in those first warm days of the spring. The endless rolling winter greyscale landscape now a vibrant carpet of emerald greens dotted with millions of flowers as blossoms drifted from the fruit trees on a gentle fragrant breeze. Alarmi's lead rope laying loosely over his neck; he is free, but chooses to stay with me. We work and ride by the stables, and I again turn him out to graze freely as we have no fences here at all. Alarmi doesn't run off, he begins grazing and looks at me as if to say

"stay a while with me".

So I sit with him, I sip slowly on my beer, watching the sun sink towards the horizon with the noise of a distance vehicle the only sound cutting through the silence. I watch as a blue sky turns first a light peach color, then amber, crimson, then grey, before the first stars begin to punch light holes in the coming night sky. Alarmi grazes beside me still; my beers empty, so I get up, stretching my legs as I a had sat way to long like this. My left leg is already half a sleep and tingles as I walk with a slight limp towards Alarmi, rub his nose and feel as unwilling to leave as he does, but Pete's made dinner... and I'd better go.

Jill jumps up on me as I enter the stables, overexcited, as to be fair she hadn't been getting the attention she demands daily. We've been here months now, and I walk in to be greeted by a pizza Pete's made in a wood fired stove, a glass of local wine and a seat at the bar he's built. We have a flush toilet we've built now too, a shower, and we've built as a great double bedroom and fixed up all the stables. Things feel... well I can't put it into words.

But I am not the same girl I was; and I know the time is coming where we will be leaving this place as Pete and I are finding it hard to constantly deal with the actions of a few of the locals. We have both put so much into this place, and they just take without any offer of anything in return. Our hosts are amazing, but the villagers... well, they are just difficult mountain folk that's all and it was really starting to take its toll, especially on Pete who is patient to the point he's not... then he becomes someone neither of us particularly like.

The next morning, Alarmi was limping, having no doubt fought with another stallion over-night and disappointment swept through me as I had hoped for one last ride before he headed back to grazing up the hill. His leg was only the start of a serious of drama; and even though we loved it here, things were being stolen from us, the locals just did whatever they wanted, and we both kind of just snapped, packed our bags and went on to Tbilisi despite desperate pleas for us to stay.

As the landscape flew past the window, it was just a blur as I stared blankly; my thoughts were still with my two boys, with the last few months playing over in my mind like a movie. All the work, the wins, the mistakes, and the last time I looked into Alarmi's eyes, the last time I touched Kashlama's Fluffy coat and rubbed his neck... I'm wiping tears, it was all just so bitter sweet.

Only a few days later; on a Wednesday at 11:58 o'clock a message came through.

"We have bad news. Alarmi got sick yesterday with tetanus symptoms, he couldn't move and was stuck in one place for the whole day. All recommendations of the vet were followed; but unfortunately, he did not survive."

There's nothing more I want to write about this...

Chapter 5.
Never alone.

"Mateship is an Australian cultural idiom that embodies equality, loyalty and friendship. Russel Ward, in The Australian Legend (1958), saw the concept as a central one to the Australian people. Mateship derives from mate, meaning friend, commonly used in Australia as an amicable form of address."
Wikipedia.

When I first sat down to write this chapter, all I could think about ways the boys I served with overseas during my time as an infantryman. We are all as close today as the day we got back from deployment, and the experiences we shared will bond us as brothers forever. I knew I would meet amazing people on our journey, and I never knew it would end up "our" journey, not my journey as the plane lifted off from Melbourne airport over a year ago.

    I have spent a lot of time on flights, with work and adventure having taken me all over the world and I usually just sink as many free beers as I can, think about the coming job and try not to think how long it will be until I see home again. This time was different; there was no return ticket booked, and I had no idea what the job would be so I just sunk as many free beers as possible and watched a few in flight movies. I cried a lot on that first leg of my Journey from Melbourne to Shanghai. I don't know why I'm admitting that... I guess I just should admit and embrace the fact I can't watch the movie War Horse without tears.

    But it wasn't the only movie I watched on the flight that made me emotional. I got punched big time in the feels by two Australian classics; The Castle, and Red Dog. If you haven't seen them then I'll sum them up for you in a word... "Mateship".

    I had no idea at the time, but I was about to meet some of the best mates I'll ever have.

Gunter.

The unknown might be the most frightening part of traveling for some, but for me it's the most exciting. Seeking new experiences, other worlds, cultures, and maybe just to understand myself a bit better... maybe leave something behind I can't handle, or just to settle the restlessness inside even for a moment. When we met Gunter, he was simply saw himself as of such an age, he had nothing left to lose but day light, and he had run out of reasons not to take his old Toyota and simply drive East. He is one of the kindest, and at the same time maddest, old German fellows I have ever met.

Our paths crossed, by luck, coincidence or maybe fate, who knows... but what followed could only be described as magical.

Looking at our saddle boxes, bags, and clothes all over the floor in our little hotel room realizing it wasn't the only mess here, we knew we needed to find a ride if we were going to get to Russia before our transit visas expired. I had hitched a ride solo lined up by the hostel I was stay at to the White Lake in central Mongolia with all of our gear, and Pete had followed in the bus the following day as he needed to wait for his Russian visa to process in Ulaanbatar. Listening to the horses pass by on the street, looking at our luggage and wondering how to cover the next few thousand kilometers wouldn't bring us any closer to Russia, or our goal of Kazakhstan.

Our empty stomachs had us wandering the wide dirt streets with their patchy grass verges and numerous tire ruts, leading us to another small hotel we'd been told cooked reasonable meals. It was a typical shambolic looking two-story hotel, with a little green backyard, again typical for Mongolia; it was full of old car parts, rubbish, building material and tall rank grasses that were clearly not palatable for the wandering stock. Parked among the mess was an old blue Toyota wagon, weirdly enough... with a German number plate.

As we sat down for a meal, we met some other travelers that were parting ways with the German owner of the Toyota, and it was suggested we ask him for a lift as he was headed towards Russia but was off hiking up a volcano nearby for the day.

We spent the day by the river together, enjoying the sun, chatting, getting to know each other and Pete took a while to teach me a few useful knots, first aid and general tricks for outdoor living. That afternoon we returned to the hotel and were greeted by Gunter, a tall, grey haired gentleman in a cable knit jumper, denim shorts, socks with sandals, sporting a huge camera around his neck. His smile was infectious, and he immediately offered us a lift in "Penelope" as she had been named. Penelope was one hell of a battler, and a credit to Toyota in every way imaginable, even if only 4 of her 5 gears were still usable.

Gunter told us proudly how his son told him she wouldn't even make it to Mongolia, and knew he will prove him wrong further still by driving her back to Germany. The numerous dents and scratches on the fading blue paint added character and charm, just like the deep furrows on her owner's weather-beaten face that always seemed embossed with a big smile. He was always looking for traveling companions, sharing story's, adventures, laughter and of course the gas bill as he wandered somewhat aimlessly in search of... well, just living each and every moment.

We left the following morning on a road trip ruled by late nights by the fire, vodka, beer, 80's music, poetry and Aussie hip hop while rolling closer to the boarder. Gunter gave me a glimpse of the home I so often wished for, reminding me of my Dad, my Uncle and at times it felt almost as though I were at home in Bavaria. The Landscape passing by was like a painting; such perfection, completely untouched and we had just the right mix of characters to share it with. Without Gunter, I don't know if the trip Pete and I were about to take would have ever worked, as Gunter gave us the perfect environment to bond, learn to trust one another and build an understanding of how our partnership may work.

Gunter was so much more than just a travel buddy; he cared about us from the very first moment we met, and then as he came to know us, we formed a little family of sorts. Gunter was a character for sure, and the way Pete and the old German interacted at first, compared to a few weeks later still brings a smile to my face.

The first days sitting for hours in the car, Gunter took over the role of my Dad, questioning if Pete's beer drinking at 3pm was healthy, and maybe there would be consequences for me as a result. Pete is a challenging character not only when you first meet him, but as you get to know him as he only gets more and more complex in his simplicity. Pete was enjoying the sun, wind, a beer and freedom that comes with sitting on the roof of a car tucked among spare tires, when I answered Gunter's "Dad like" questions about Pete, knowing it was out of concern for me, and that he had nothing against Pete at all.

"That's on him, he is well aware he drinks too much but doesn't harm anyone, he just loves a beer and a laugh"

We didn't agree on that one but he accepted my attitude at the time, what I didn't know then were the side effects of hanging out with Pete. He doesn't try to be someone he's not, at least not out here, and just as with horses, Pete doesn't judge people, but takes them for who they are. Gunter started to really relax as a result and started to swear and drink more, it was more than that though, he just started to feel more at peace as each day passed and was soon laughing from the roof top with a can in hand as I drove the boys. The same happened to me, and arriving at our last camp before we left Mongolia I was torn apart at the thought of soon parting from Gunter, Penelope, and our little family.

But Gunter left me with a gift the day we parted that still has me laughing any time I think back to it. On arrival in Kazakhstan, we grabbed out some cash from an ATM, and as usual Pete used his currency converter app and offered it to both Gunter and I. Gunter refused, saying he had a grip on the conversion rate, so Pete and I left him to it...

A few moments later, I lay in the street in tears laughing. Not only had Gunter purchased 5 kgs of plums that would soon be spread through most of the car, but he'd missed a zero on the conversion rate; he was now holding enough Kazakh currency to purchase a house. He just stood there eating plums one after another with a goofy grin on his face, sharing a bowl of cognac with Pete as though it wasn't a big deal.

His plum-stained beard and now somewhat confused look just made it worse, I completely lost all control of my laughter and began to panic I may in fact wet myself laughing right there in the street. Pete decided to film me, attempting to get me to re-tell the story through my tears for both his, Gunter's, and the local street vendors' amusement. It wasn't my finest moment I'll admit, but for me, moments like this, with guys like these, make traveling a life worth living. You just never know what the day will bring, but when you have a "Gunter and a Penelope", every day is going to be one filled with laughter... and sometimes the smell of fermenting under-seat plums!

And they say Germans don't have a sense of humor.

Horsemen of the steppe.

I tucked the hood of my Swazi coat a little tighter around what little of my face was exposed, shuffling inside my layered merino, puffer and Swazi protecting me from the light evening breeze that I'd estimate at somewhere around -20 degrees Celsius. I'd just cleared a long patch of low brush that had kept both Billy and I a few degrees warmer. We paused, and I struggled to look back toward Luisa, Jill and Jac due to the restrictive layers of clothing, opting to just turn Billy in a slow circle instead. She'd taken a slightly different path to me, and was some 100 meters back from us now.

Billy swung around without a cue from me, turning his backside into the breeze as he began to graze while he waited for his mate, leaving me to ponder the last conversation I'd had with a fellow Australian a few months earlier. We were in the North Western most corner of Mongolia, and were chatting in the markets just before we crossed into Russia heading for our intended start point of Pavlodar Kazakhstan…

"I'm in the S#$%, ha ha. Me Mrs is off me, been hit by a car this morning, and I had to have four shots of vodka with the Russians... And now I have to find the tools to fix me F*&$#n bike, cos I broke it last night when I was driving pissed."

That was the last conversation "word for word" I'd had with a fellow English speaker. Luisa had traveled with him and two other Aussies briefly during her long stay in Mongolia, and like us, they too were all a bit mad, traveling as a group of three on two cheap Chinese motorbikes covered in gear. They were all covered in tattoos and piercings, and although nothing like us in many regards, their free-spirited nature and lust for adventure was exactly the same as ours. I wondered where they were now, and wondered what my mates at home were up to and how the fishing had been going of late.

There's lots of time to ponder things like this on the steppe, but sooner or later I was going to need to start considering where we would stay tonight as the temperature was literally dropping by the second; and the sun, obscured by pale grey cloud rolling from horizon to horizon, would soon be gone.

We had cleared the scrub of a small rise and bellow us, a small river separated us from a sprawling compound a few hundred meters ahead on the edge of a huge flat basin hemmed in by mountains some 5km or so in the distance.

After a brief chat, we both agreed to make for the mountains, and camp in the shelter of some distant rocks and scrub; but first we would water the horses and stop by the compound to see if we could obtain a bit of extra feed for our boys. As we entered the compound on foot, leading our horses we were immediately greeted by as usual, confused, but very welcoming and friendly Kazakhs. We did our best with google translate, and it was soon clear we would be welcomed as guests whether we liked it or not. The next few days had grim weather and freezing temperatures on the way, and our horses were to be welcomed into the barn and fed up before we set of again.

Now think about this for a second... two people you don't know, who don't speak a word of your language just wander up to your front door. Filthy I might add, with a dog and two wild horses and ask for hay and water using pictures on a cell phone. What would you do?

Luisa and I were starting to understand just how different the people of Kazakhstan were from those at home. Strangers are so often viewed with distrust and fear at home, yet here we we're treated as though we were family. I would say cousins, as much as anything else... as though we were known through mutual family members but only met over Christmas or other such holidays. Our arrival was treated as such, and we were soon enjoying vodka, delicious sliced horse meat and a stunning spread of food the likes of which I'd never seen before. Neither of us could recognize much of it, but the tastes were amazing and the warmth of the place!

It was over 30 degrees in utter contrast to what we had been braving the last few days outside on the steppe. Luisa's "cricket smile" couldn't be wiped from her face that evening, or over the coming days we spent exploring the local area with our hosts. They had a huge leased area of land where they had well over 3000 horses, hundreds of cattle, sheep and also bred race horses. We rode out the worst of the freezing weather in their compound; sharing stories, pictures, and learning about their farming practices as our horses enjoyed the luxury of a barn, grain and lucerne hay.

Moving on was done with very heavy hearts, and we were given many gifts we will treasure for a lifetime; none more precious than the way the Kazakh's treated us as though we were kin. But this was not a rare event; this was literally the standard in Kazakhstan and not just each evening when we looked for supplementary feed for the horses. Word traveled ahead of us via WhatsApp and Facebook, and whenever within sight of the road we would get visitors literally driving into the steppe or waving us down to take photos with us and give us beer, vodka, fresh food and even cash!

The further south we pressed, the colder it got, and the warmer the reception was with people more often than not everything short of arguing over who would have the chance to host us for the evening! I literally can't make it up, the police would stop, put on their lights and walk out onto the steppe for photos with us before parting with smiles, handshakes and a gift. Road workers would literally bring us hot lunches in boxes, thermoses filled with coffee and tea and invite us into their work yards to eat with them!

It's very hard to put into words... the way we were treated by these amazing humans and horsemen of the steppe. I just can't explain the warmth of a smile from a toothless weather beaten face on the freezing wind swept steppe, as he is honored to hand you a bottle of vodka and some cash to shout you a meal as his guest, in his nation, that he is so distinctly proud to represent with his kindness and warmth. It's almost like the colder a place is during winter, the warmer the hearts of its people are.

In the movie "Borat", which in many regards mocks Kazakhstan, there is a line in the Kazakh national anthem made for the movie stating "Kazakhstan is the greatest country in the world". That line is the closest thing in that whole damn movie you'll see to an accurate depiction of how we have come to view both the people, horses, and landscape of Kazakhstan. I would without a hint of hesitation recommend visiting this incredible place as it's filled to the brim with some of the kindest people on earth, even if it is bloody freezing!

Uzbekistan, a word I will always associate with love.

I met Jahangir over the couch surfing website, and it was through him the experience of a life time opened up to Pete and I. Without him... who knows? But I doubt my life would have become what it is today.

Central Asia is ruled by horse people; by wild, drunken men. Life here is hot and dry in summer, and a sub zero, windswept wasteland in winter. A hard, rough and old fashion labyrinth for a Bavarian woman and her dog, but I was never alone, nor felt alone either. After escaping the dinner table for a few minutes to freshen up in the bathroom, I gaze at the reflection staring back at me; taking my mind back some two years. I stare at my shaved head, a decision I don't regret for a second as I recall looking at my beloved hair neatly tied together with a back piece of string on the table beside the couch.

I was always too scared to do anything with it, fearing I would look stupid... and then, I just shaved my head and sold my hair. The first time I looked at than new person in the mirror, she seemed so familiar, her eyes sparking, full of life, almost as if she has just been freed from a long captivity in societies forced normality. I splashed water on my face again, smiling seeing and feeling that same freedom in my eyes as the days long since past when I first got rid of my long mousey blonde hair.

I know I am expected back at the table, where no doubt another full glass of vodka is already waiting for me. Despite Pete's comfort around these people, the way he moves in their land, understands them after all he has suffered at the hands of both the cruelty of life and the environment much like the hard men of the steppe, I still stumbled like a drunk; smiled at, but often more so tolerated than accepted.

It wasn't until we met Sabir that I started to feel as though I fitted right in, finally seen as a horseman first, regardless of my being female. Sabir was no taller than Pete but had a good ten kilos on him; like Pete he was bald, with a weathered face and eyes of piercing intensity. Clearly a powerful man with skills honed over 50 years of hardship on the Uzbek steppe, he entered our hearts the moment we saw him interact with his horses. They were monsters; massive Karibair stallions who just oozed intimidating power from their near perfect builds.

But that wasn't what really fascinated me, it was the way they looked at him, the way he touched them and the way they talked to each other. They were lambs, and every touch, look and movement was spoken with a softness and kindness that defied all logic if Sabir and his horses were to be judged on appearance. I was in the presence of a true master horseman, and he respected me!

Sabir's horse's where the first we looked at after entering Uzbekistan, and we knew at once we could never look at a horse the same way ever again. I recall the second we drove onto his property the conversation in the car suddenly stopped, as both Pete and I looked at each other as if to say

"only in our dreams can we afford these horses."

I nearly didn't notice the friendly hand offered to me as I exited the car, transfixed by the pure beauty and strength of the two stallions standing beside a small stone stables among fruit trees and vegetable gardens of a house on another unremarkable street on the edge of a sprawling city. Financially, Sabir's horses were of course out of reach for us, but we enjoyed our visit, unaware it would lead to more than we could have ever imagined. With Sabir on our side we entered the world of Kupkari; a game hidden from the outside world, hidden from all who couldn't possibly understand the sheer madness and purity of its origins in mounted combat.

Music filled the air; stallions squeeze through the endless streams of cars and people heading towards the freshly harvested wheat field on the outskirts of the city. The smell of cooking fires and barbecued meat hung in the air as man and beast alike moved enthusiastically towards the celebration that lay ahead. Stallions; they were all stallions, noses flared, heads high and moving like strutting studs ready to fight to the death over a lone mare. Their riders poised, expressionless, at one with their horses as they moved confidently in complete control among the jostling and cheering crowds on foot.

I'm having flash backs as I type, I can't stop smiling... the smells, the sounds, I wish I could find the words in either my native German or English but they simply don't exist!

I was the only woman on the field, though many others were part of this amazing spectacle; but they are the ones turning the Shaslik over the red glowing coals of grape vine fires, the ones filling white Plastic cups with vodka and the cheering men on from a safe distance. But I; I was among the warriors and stallions! A little Bavarian woman with a shaved head and crooked smile I couldn't wipe from my face.

Word had got out we were present, and the organizers welcomed us, announcing over the loud speaker we had guests that rode here on horseback! Not even 10 minutes later, Pete was being lifted onto a horse so hot even he appeared to flinch as it reared and fizzed, engulfed by the cheering crowd. In a second, he disappeared at the canter among hundreds of horses and riders into a game we still struggle to comprehend the rules of!

There was nothing more than a field with three circles made of hay before us, seemingly no boundaries, and a crowd that ebbed and flowed across the field like waves on the beach as the thundering wall of horses plowed across the field with no regard for anyone or anything's safety! I'm the only woman here, grinning from ear to ear with a policeman and Jahangir standing beside me for protection tingling from head to toe! This is utterly unbelievable, so much so, and with the sun in my eyes I didn't recognized Sabir coming towards me, leading a slightly smaller grey horse by his side. He smiles, leans from his towering stallion and hands me the reins...

With everyone expect Sabir watching on critically, I couldn't be more proud and thankful. It was so much more than just the riding of a Kupkari stallion, it was about the trust given to me, and the statement made by these amazing people opening up a door into a world I thought I would never have seen, let alone find a place in.

Days later, still filled with the overwhelming feelings from last weekend's event, we made our choices regarding horses, though a bit foolish on reflection in part as much due to our optimism as financial limitations. The pressure of choosing a horse at first sight, the expectation of the people watching, the visa's limited time constantly ticking away... it put us in a hard spot and moves simply had to be made.

So, as we rode off into a snow storm, Pete was sitting on a ticking time bomb, with the act of simply getting on or off Beersheba a drama itself. Jill happily running around jumping up and down next to me didn't help me in the task being laid squarely on Bluey and me. We had to lead Pete and his boy safely behind us, providing the rock on which all Beersheba's drama must break upon. We did lead... into one disaster after another and with each close call we both grew immensely as horsemen and riders.

With that first snowstorm and a few days passing, Pete found huge joy riding the little hand grenade with the pin out, the two soon mischievously playing up as much as each other while I still sat rigid in consistent, steady lead of my boys.

But the snow picked up again and Bluey ended up struggling with the cold. As impressive and strong the Kupkari stallions are, they are used to warm stables, blankets, grain and even warm buckets of water. My mood became as frosty as the ground beneath us after a few nights dropping well below zero, made worse by the minute with Beersheba and Jill being idiots. Jill took a run at Beersheba as Pete dismounted, dislocating his shoulder managing the young stallion as Jill braked and pushed while Pete soothed and calmed. I felt I was breaking under the weight on my shoulders, unable to control the tears, but proud enough to go for walks so not to be seen in tears by my partner whose pain was taking a toll on his mood too.

Look, I love Pete; but the truth is his body and head resemble a vase that's been smashed, and glued back together by a child with crazy glue.

My pride nearly stopped me from calling the Jahangir, the man who had already done so many favors for us, but I had no choice. Jahangir, a man so rare in this world; his heart was wide open, willing to go the extra mile without expecting anything in return but the pride that comes in knowing you daily make the world a more beautiful place for everyone in it. We'd started chatting when we were still in Kazakhstan, and he'd since taken us in, introduced us to Sabir, as well as patiently driven us around to find our horses. He would take us out for dinner, proudly introducing us to local cuisine and offered us a warm comfortable place to stay. The truth was, I just didn't know what else to do; and as soon as his voice appeared on the other end of the phone... it felt like a soft hand placed upon my shoulder.

A veterinarian was on his way just a few hours later, and he even got us horse blankets delivered the next morning. There was no drama for him, just friends in need and the very moment he picked up the phone, he left everything behind to help us out.

I thought it took years to build up friendships... you know, the real ones that last a life time. But here, a little bit lost in a country still so strange to me, with the wind howling outside I sit with a little bowl of fresh milk warming my hands, wrapped up in a colorful hand stitched blanket.

We reset, and started again; and for a couple of days it nearly felt like being back in Kazakhstan, camping on riverbanks, drinking beer and playing cards while listening to the horses happily chewing their hay and grain rugged each night in nice warm woolen blankets. Those first days of Winter changed everything; and it was as the first real snows and frosts that didn't lift hit, that we left the safety of the city and pushed into wide open spaces of the Uzbek steppe. But it was nothing like the empty steppe of Kazakhstan, more a maze of deep, wide man-made canals and irrigated farmland sprawling from horizon to horizon between cities.

We had to reorganize our routes and riding accordingly. Shepherd's on horseback's friendly waves from a distance and empty ghostly streets of the small Kazak villages were now replaced by big towns, each with a new challenge for Beersheba and Pete.

My biggest worry though was Bluey, even though he'd been going a bit better for the last couple of days, he soon felt really limp again. He still struggled with the cold, and he had started to show a bit of lameness in his left rear leg as well. The invitation of Jahangir "once again" to stop at his family's home not far from our last night's camp seemed like a slice of heaven, especially as he had a small stable ready to rest the horses out back, and a warm bed and shower for us! Between us and him, only a small city separated us on the map.

Well, we misread Google maps, and misjudged our capabilities, and decided that with a lame horse we would push right through the middle, the shortest way to the beckoning warmth of Jahangir's home.

Soon, we found ourselves running out of room to ride, and endless obstacles hemmed us into a terrifying gauntlet of bridges and motorways with narrow or non-existent shoulders. But we were past the point of no return, it would take days to back out and go around, or hours to push through. The street we now found ourself on had turned into a highway without any warning at all, and it had disaster written all over it.

Pushing Bluey up was causing me physical pain, feeling him carrying his leg so I stop, get off and tell Pete I can't do this any longer. I start walking as cars fly past us at 100kph, I'm scared to death, and I hear Beersheba behind me starting to lose it. I look back, he's trying back into traffic as Pete holds him calmly, aiming him at the guard rail and pushing him forwards, managing him without the rock Bluey and I provide for him while mounted. I put my hand apologetically on Bluey's neck, knowing full well that me being on Bluey doing my job as a lead horse could be todays difference of life or death for Pete.

I swing myself back in the saddle, hardly recognizing Pete's thankful words as his horse visually starts to relax again. Jill seemed to pick up the vibes, tries to disappear under Bluey's tail with timed submissive look. Pete keeps talking at the back, his motivational speech didn't seem to come through to me at all today; all I can think is just one more step, one more. I picture Jahangir's face, like a peaceful port in a storm and keep telling myself... we are nearly there.

The narrow shoulder on the side of the road disappeared as we entered town, and we now need to choose between racing cars, and a small foot path in front of sad looking single-story houses standing side by side. The elderly folk sat quietly drinking and playing card games at the house front's caused Beersheba further distress, so the only way for us to go was the road. Trees lined the street, all seeming to bend under the weight of the fruits hanging from them. They creaked in the gentle breeze of a beautiful warm cloudless day, almost like the empty blue sky above was laughing at us, the weather we wished for the past few days appearing exactly when we couldn't care less, knowing we would be in a warm bed shortly.

But we were not there yet, and ahead lay Beersheba's pet hate... a vehicle taking up the entire lane with mechanics removing a wheel and busying themselves with its repair. Beersheba trusts Pete, but absolutely hates men in general and if one so much as goes near his face he, will back away regardless of what's behind him unless pushed forwards ruthlessly by Pete. But if I trot, Beersheba will almost always follow so long as Pete can stay in tow.

I want to throw up looking at the truck parked ahead knowing I may only have one shot to drag Beersheba past it. With every step brining us closer, I feel more and more sick; stopping right in front, looking back at Pete who nods, urging me to trot. I check a last time for cars, and go for it. There's just so much noise, and I scream Jill's name to follow and she does, but Pete seems to struggle. Beersheba's freaking out under him, with the cars here so used to horses and donkey carts on the road they cleared him by centimeters; none even bothering to slow down.

My eyes opened wide in shock, a scream stuck in my throat as the cars pass Pete and Beersheba missing them only by cm as he manages to keep it all together with the poise of a seasoned Kupkari rider. It was though Beersheba's fears were fed by my apprehension and shock, and in desperation Pete yelled...

"KEEP MOVING!"

I go, pushing hard; too hard, feeling the pain increasing for Bluey with the burst of speed I start shaking, somehow knowing everything is going to be okay. I look back, sinking into my saddle in relief as a settled and calm Beersheba trots up behind me... they mad it.

We don't talk, we just keep going, one step after another with the cars passing close by. But we made it, and with every single incident like this Beersheba quietened more and more. As my brain starts to process what just happened, I drift off, and just try not to think. You can do it Lu, it will be over soon and we will sit with Jahangir having dinner, laughing about what just happened in just a few more steps.

That never ending road through town, the pain and fear I felt operating in a state where I had nothing more to give, it allowed me to grow like never before or since. And I was right, sitting later that evening at the kitchen table, still exhausted but smiling, sharing the past few weeks with our friends made me feel better in an instant.

The warm soup, made with so much love by his Mother made me feel at home, blessed, supported and surrounded by strength; but most importantly loved. Loved and complete in the company of life long friend's I'd literally only just met.

There will always be those that seek to spread hate towards those of the Islamic faith and ridicule them saying "it's a religion of peace". Well, in our time with these people, peace and love was all we felt, were shown, and saw. Sure, a woman's place is different somewhat to that in the west, but both of us feel that the love and respect we were shown through our time in these lands was as pure as the snow swept steppe in the dead of winter.

We saw a religion of peace, and they have nothing but our love, admiration and respect regardless of faith. Jahangir, Sabir, their families, and all those who helped us without so much as a question of personal reward during our time in Uzbekistan are simply a credit to the human race, their country, and their religion.

New lands, new challenges.

Azerbaijan felt different to the Stans from the moment we got there. Cashed up on oil money, the capital Baku moved at a different pace and although friendly in its own right, it felt more like the West. How so? Well people were nice, but it seemed they were nice for reward rather than just the sake of being nice. And that's fair enough, people have to eat, people shouldn't be expected to do a dam thing for free and I am in no way speaking negatively about the place. It was beautiful, warm and progressive with a hell of a lot going for it.

    We met great people right from the moment we got there but just didn't feel the meaningful connections we had with the Mongols, Kazakhs and Uzbeks. Maybe if we had longer there, things would have been different, and the day we arrived it looked a great deal like World War III was about to kick off with the Iranian's accidentally shooting down an airliner. It was these tensions and other regional issues that had seen us cross the Caspian Sea rather than crossing through both Turkmenistan and Iran despite having visas for both.

    It's sad really; I'm sure we will be back these ways at some point and we had such wonderful contacts in both countries that were willing to help us along the way, and just the thought of riding through ancient Persia... wow!

    We made poor choices again in Azerbaijan, choosing to give a donkey cart a try as we were having a very hard time finding suitable horses. It was costing us a fortune to do nothing, and the plans we had made kept stumbling with donkeys being both slow and difficult. After the freedom of the steppe on horseback, a donkey cart was like a road trip with your parents in a diesel van with no stereo, having just stepped out of a tuned sports car.

It's a shame, the borders and the difficulties getting horses across meant we needed to change from horses we knew, too searching, and we just didn't have the support to look long enough with people that knew the country and for that matter horses. The other problem we faced of course was "tourist pricing". Sadly, you're always going to face that without a local whose heart is squarely in the right place, so with growing frustrations, a rapidly shrinking budget and border tensions increasing towards the coming war with Armenia, we pushed onto Georgia.

I'd been to Georgia before and absolutely loved every minute I spent there. Here people were welcoming, it was affordable, and I'd had a chance to ride with some amazing people whom I'd got back in touch with about horses.

They were genuine, I knew they had great horses and really enjoyed their attitude to horsemanship during a brief day long visit back in late 2018. We'd had a great day's riding together, and the food, wine and hospitality they extended to me was utterly exemplary and world class in every regard. Tatia had been running her small horse trekking business for a while knowing full well it had huge potential with a bit of extra input, and she just welcomed Luisa and I in with open arms during the cold of winter to break horses for her and upgrade the facilities at the stables.

It was a win-win situation in every sense of the word for us both; and although we rarely had much contact with Tatia, it was through her kindness and that of her partner that we met many amazing people during our stay in the amazing nation of Georgia. Below the stables in that small village that Luisa and I came to call home for many months lay another small settlement made up primarily of summer homes for residents of the nearby capital Tbilisi. It was here we met Guga and his family and were shown the heart and soul of the real Georgia, flaws and all.

Guga is a man Luisa and I have come to consider extended family, as that, like so many others before, him is how he treated us from the moment we met. Guga is as genuine a man as you will ever meet and a true Georgian patriot. We spent many a night discussing Georgia's past, its future and its issues in a rapidly changing world, and unlike so many people we've met along our path, Guga has an incredible grasp on the issues that hold a place like this back after time in both politics, the private sector, and as a member of both rural communities and big city life.

It was this wealth of knowledge and want for nothing more than good friends to surround himself with, that we found ourselves coming to trust another character who's help, like Gunters, Jahangirs, Sabirs and Tatias before him would steer both Luisa and I in a new direction.

Guga understood our frustrations with village life all too well, and alongside Tatia they both helped us no end when the time came for us to move on. They both, for no reason other than having generous hearts full of love for others and their nation of which both are immensely proud helped introduce us to others whom which we could volunteer our services for in exchange for a place to rest our saddles and bags while awaiting the end of the COVID lock downs and boarder restrictions.

These people... they are all the salt of the earth; and as westerners experiencing these cultural differences here in these former soviet republics it has done nothing short of take my breath away as to how kind, beautiful and caring these people can be. Most have so little by western standards, but everything they have they share knowing its relationships that enrich our lives, not money, possessions or the job title you hold. I've heard the saying...

"He who dies with the most toys wins."

I can tell you by this definition Guga could be considered by some as poor. But our lives have been enriched beyond measure by every minute we have shared, and will share in years to come. Guga is the perfect example for me that true wealth comes not from money, but from sharing your love, laughter and life with others.

An alter boy and a ballerina.

The Tusheti region of Georgia is a place I can only describe as how I pictured ancient Gaul reading Asterix comics as a kid. It is truly wild, and the drive in is in parts terrifying. While we were there, a mini bus went off a cliff tragically killing 17 of its 19 occupants, and on almost every road or horse trail there are sections where you can fall as much as 1000 meters before finally hitting the river below. It's breathtaking in every sense of the word, from its forests, fortified towns, spectacular cliffs and alpine pastures, it's a wilderness like nothing I have seen before. In parts, it almost reminded me of the South West coast of New Zealand and its plunging forested gorges melting into snow grass tops; though here the brown snow grass was replaced by summer fields of wild herbs, berries and lush summer grasses of the most brilliant green hues.

Every part of it seemed to hold a story, and it was here the Georgians had fought off many an invading force through both skillful use of terrain, fortifications, and the brutal winters that see these remote alpine villages locked in by winters snows for more than half the year. The stories of those that lived here are often passed down through songs, and it was these songs that first brought John, a native of New Mexico, to Georgia back in the 90's while studying fine art in Russia. The songs themselves are amazing, but it's the stories they tell that I find particularly moving.

For example, there's songs about the ride to the low lands in spring to procure salt for the preservation of meat and other foods in order for villages to survive the harsh and completely isolated winter ahead. The journey was always potentially deadly, with raiders, bandits, invading armies, let alone the environmental hazards; and hearing a song sung by a father to his son as he embarks on his first horseback journey into uncertainty is deeply moving.

Regardless of the language barrier, hearing someone sing about finding their son's shepherd stick among an untended flock, realising their son will never be coming home again is deeply emotional and moving. I understand why John was drawn to this culture, and have merely glimpsed his understanding of it, but now too feel drawn to these amazing people and culture.

We were introduced to John through Guga and Tatia, and came to stay at John's ranch in the fortified city of Sighnagi in the Kakheti region of Georgia. John is a sort of "organic and spiritual prophet" among many other things. An artist, restaurant owner, chef, musician and organic wine maker, I found our time together in both Kakheti and Tusheti life changing as many of my preconceived ideas and views were not only challenged respectfully, but in many cases forever changed for the better.

It would also be wrong not to mention John put me in touch with a friend of his, a cardiologist, that in less than an hour noted and treated a genetic heart condition I had been complaining about to doctors for some five years without a correct diagnosis. It sounds dramatic, but in doing so he saved my life, at best saved open-heart surgery anyway. But this is a book about horses right, not organic wine making and heart issues? Well the story must be told in its entirety and it was through a mutual love of horses that all these doors opened to these incredible experiences Luisa and I had together.

It was at the tail end of the Georgian summer when I first visited Tusheti with John and a few of his staff from his Tbilisi based wine bar "Vino Underground", and restaurant "Poliphonia" in search of wild foods and herbs to forage for on horse-back deep in the mountains. It was quite the journey even just getting to our start point, some four hours of winding switch backs on a gravel road that fell away literally into empty space bellow. At first we contoured along the side of a deep forested gorge lined with memorial crosses at intervals of no more than 100m, marking the places many an unwary or unlucky motorist had fallen to their tragic end in this life.

As we rose higher in elevation, the forested cliff faces and jutting ridges rolled into stunning mountain top pastures interrupted only by splashes of color from wild herbs, flowers and the scattered granite of the crumbling high peaks above dragged lower by the movement of winter snow drifts long since melted. I was at any moment expecting to see Gandalf ride past on Shadowfax, Gollum lurking in a rock crevice or a couple of little hobbits scamper past as we crossed the high pass marked with yet another memorial cross before entering the heart of Tusheti.

When the road ended, we met a string of horses, three of which I later learned all shared the same name whose English translation means "orphan"; all had lost their mothers to wolves.

I was to be riding one of the orphans; a stunning off-white colored stallion born and raised in these mountains, and to this day he is without a shadow of a doubt one of the finest horses I have ever had the pleasure of working with. As we saddled up, and loaded gear onto the pack horses I was excited for the coming adventure sure, but more so who I would be sharing it with... artists, wine makers, a brewer, chefs, and the wild men and women of Tusheti itself.

I have hunted my whole adult life, so heading into the wilderness feels right, but this time things felt different as I wasn't thinking about the ballistic calculations I'd need to make on the fly for a long cross valley shot at this altitude of some 2200 meters above sea level... I was thinking about the first conversation I had with John months earlier about wine.

I had tried an organic wine variety called "Kisi" for the first time and remarked on its remarkable flavor and subtle tones... canned peaches, a hint something I couldn't put my finger on too, almost like incense burning in another room. John stopped me, suggestion I was tasting the wine sure, but pressed me as to what I felt when drinking it. What emotion did it remind me of, a feeling, a song perhaps? I was dumbfounded. I heard the Beatles, I heard it creeping in from long forgotten corners of my mind as I walked down the long corridor separating my brother and my end of the house from my father's office where he worked as an author writing natural history books.

In that one moment, John changed the entire way I viewed wine, something it's fair to say I am very fond of and shared a passion for with my late grandfather who loved wine so much, he had to move house so he had more room to store his collection. From that moment on, the world of organic wine and food began to open to me in ways I am still trying to understand from a scientific standpoint, as I now have no problem at all in noticing the difference between a glass of wine Luisa has held, and one I have.

But it wasn't just the wine, it was John's attitude and philosophy towards food, its production and preparation that blew me away, as best as I can I will try and explain his thoughts briefly in his words.

"When it comes to meat, or the production of any food or wine; if you buy rubbish then you are a supporting a system that is counter-productive to the health of the world and for that matter yourself. It's not about being vegan, or that eating meat is bad, and organic is the only solution; it's about supporting ethically sourced produce that has as little negative impact on the earth as is possible, and as positive and impact as possible on the communities that produce it.

It's better to eat a piece of ethically produced free range wild meat, than it is to eat a Monsanto produced genetically altered carrot drenched in pesticides throughout its life; and vice versa, if you can't find a good piece of meat; settle for a carrot that was grown with love. Natural wine is a form of art, and it has a certain asymmetry from bottle to bottle which is part of its beauty; much like the eating of a wild oyster, you don't expect each one to taste exactly the same. This variability in organic wine and food is part of the joy, journey, and adventure of the experience.

Where if you buy Doritos, Mc Donald's or coke then your consuming a laboratory creation that has been regulated with chemicals and additives that basically make it nothing more than highly addictive junk food for people who don't know better. It doesn't mean it shouldn't exist, and it doesn't mean someone is morally better or not by eating it; but what it does mean is if you choose to eat it you should take accountability for what you're doing to your body and to the environment through its consumption.

It's not that it doesn't have the right to exist, there are a lot of bad things in the world that exist, many worse than Doritos; but when you purchase these items you choose to support an industry in a way not unlike that when you support a political party. You can choose to grow and purchase items grown with love in a sustainable manner, or you can choose to support mass produced foods high in chemicals grown in mono cultures, then highly processed and packaged. In the last few years, people have been voting for a lot of crappy politicians and eating a lot of crappy food."

I was coming to see John's organic wines we're nothing like those mass produced chemically stabilized wines produced in the heavily fertilized, irrigated, and chemically managed fields back home in Australia; they were alive. I was beginning to understand the importance of our food being alive in a way I had never considered before too, and the idea that since the stone age we had been creating conditions in which bacteria and yeast would create for us literally life's finest foods through fermentation in a clean environment.

All the best things are fermented; coffee, cheeses, chocolate, yogurt, beer, wine, spirits, pickles are all created through these various processes of fermentation. And it was thoughts like this that spun circles in my mind as my little white stallion trotted without a care along a trail no wider than a cell phone with a 300-meter fall one step to the left. This ride, the people I have met and been influenced by are making me a more balanced, connected and spiritual individual who is seeing things I had never seen before despite, having them right in front of me all along; and it is without a shadow of a doubt helping me to connect with horses in new and exciting ways.

The days that followed in Tusheti were full of stories, wine, chacha and trips into the forest foraging. I learned a lot about mushrooms and although they still aren't my favorite thing to eat in the forest, the sheer variety of edible mushrooms I had spent years walking past as I searched for deer and other game was a real shock to me.

But it was the herbs that changed me more than anything. I had always viewed herbal tea as something drunk by the hippie type folk; you know, the guy writing his novel in some hipster cafe with his boat shoes, no socks and corduroy pants ending mid-calf. The typical op shop cardigan, a man bun and acid jazz playing as he sips with his pinkie finger in mid-air sort of type. But I was wrong, just as I was wrong about stinging nettles. I had cursed them for some 22 years as a hunter, having to recover fallen deer from valleys thick with the stuff whose sting carries the same painful toxin as the European wasp! But nettle, these herbs that were seemingly everywhere; wow, ahh... they taste amazing!

These herbs are a beautiful gift from nature I had never even noticed. A pinch of nettle, some saint John's Wart, black thyme and a single wild mint leaf in hot water beats any hot drink I've ever had and the stuff is just growing wild everywhere! I feel ashamed at how narrow minded I have been at times, although I still think angry vegans yelling at people in restaurants have attitudes that are even worse than mine was. But it was my perception of these angry vegans that had built walls in my mind between me and the place I now found myself experiencing fully, not just hunting in. It took a gentle, spiritual guide like John, a former alter boy, and a wine bar manager who once wore a tutu too help me find true balance.

My time riding and chatting with these amazing people has changed me for the better, spiritually and emotionally. I drink less, I laugh more. I am connecting more and more with the world around me and I have horses to thank for it all.

Horses; Like John, Guga, Jahangir, Sabir, and everyone we have mentioned or had present in our minds as we wrote this chapter have been guides opening doors to worlds we could never have imagined or seen ourselves visiting.

Love is a word that can be miss used easily; but there is no other way to describe the way we feel about those that have shaped and enriched every minute of Luisa and my journey together.

Much love to you all, and thank you.

Chapter 6.
Becoming a horseman.

*"Australia has a legend that through her sons lives on.
Born amidst the colonies; a horseman thereupon.
He derives from hardy stockmen; a yeoman cavalier.
He's been tested under fire; a man with little fear.
To most countries he has traveled; through many wars he's been.
He's by far the strangest soldier the world has ever seen.
He's a rider; he's a fighter; a rouge; a gentleman.
He's the devil with mobility; a fearless guardian.
With loyalty beyond reproach that one should never spurn.
The man's respect is not a thing that you may easily earn.
He's always unpredictable and lives the life of whim.
He doesn't take it seriously; it's all a game to him.
He's a reckless man, a braggart, but in truth he's earned his due.
There is little that he says that he can't ever do.
While the time has changed the horseman's job to a different art.
And years have passed since last they charged;
he's still the same at heart.
Even though his steed has altered, he knows when things get grim.
The spirt of the Light Horseman will always ride with him."*
Poem by Ian Coate.

Luisa and I both like the term "horseman". We also like the fact that in pretty much every level of equestrian sport there is equality between men and women; with your ability to connect and work with a 500kg animal being what counts, regardless of gender. Luisa and I are both very good at what we do, but compared to those who have lived and worked with horses their whole lives, we are still apprentices at best. They say it takes 10,000 hours to become a master in your trade, but with horses you never stop learning. Both Luisa and I have been immensely fortunate in the way we have been trained not by one individual with 10,000 hours under their belt, but dozens of masters of both genders from many different backgrounds and disciplines.

It's been an amazing journey to take individually, and in partnership as we have the ability to put in to practice and discuss lessons learned individually, together. Luisa and I both agree she is a considerably better rider than me in a technical sense; she has classical training behind her, and a very solid head on her shoulders. I ride very differently to Luisa, and have ease getting horses to do things they often don't want too or wont for her, mainly due to my greater experience with wild horses. We ride so differently, but together the lessons we have learned on the steppe with our own horses have been the greatest teacher of all.

First and foremost, I must state that the way people ride on the steppe varies immensely, just as it does in the west and it's totally unreasonable to think our experiences are representative of the entire countries we speak of in this chapter and the lessons we learnted there. Mongolia alone is the 13th largest country on earth, and the way people ride throughout it varies; and due to its size, it is fair to say there are both excellent, and terrible horsemen there.

The lessons and experiences we speak of in this chapter may be found a little challenging by some of the more traditional riders reading this. During our journey together, and as individuals we have experienced things as they have been conducted for thousands of years and these practices may be deemed barbaric and backwards by some readers. For example, after a week-long ride hunting wolves in which we covered several hundred kilometers, the Mongolians practiced "bloodletting" on many of the horses including mine. I in no way what so ever support this, particularly given it was done with a dull pocket knife; but to gloss over facts like this is to not fully prepare people for what you may experience on a long ride.

In Georgia there are literally no equine vets, and the one that has some idea around horses is an alcoholic of such a spectacular degree, his work is at best hit or miss. Tragically, as a result of this I found myself in a position in which I was forced to euthanize a horse personally, as well as conduct numerous veterinarian specific medical procedures. It is not our intention to offend, but merely allow a glimpse into our experiences, some of which have been among literally the first horse cultures on earth to domesticate horses.

It has been both and honor and a privilege to learn from and ride with these men and women, and in doing so, we experienced a glimpse of another world. A world that simply doesn't exist in the west, where money buys what you want, and even the poor are fat. Here the riders are as wild and as hard as both the winters, and the horses that roam the steppe. We opened ourselves up to every aspect of the culture here, both good and bad, without judgment of them at the time and it has been for this reason we have had such an amazing variety of experiences usually "off limits" to other travelers and completely off limits to tourists.

With hindsight there are many aspects of their treatment of animals we do not in any way support; but there are also many lessons which have us questioning the behavior of many riders from the west, that would judge these people harshly.

We know we will face judgment at the end of this chapter, and we have chosen to write it anyway; as we both feel it is important to hold a mirror up to those that wish to so quickly judge others, especially online. We have seen horsemen that are just way too hard on their horses, particularly in Georgia, where they are just so hard on their mouths.

But at home we have seen people that are just way too soft, resulting in a horse that's overweight, terrified of everything, and dangerous as a result. Finding that middle ground will never happen, and no matter where you stand you will face opposition; but we have the honesty and integrity to say how we feel knowing full well we try every single day of our lives to grow as horsemen. These are just some of the lessons we feel have been most valuable to us both.

Unlearning, relearning, or just finding another way?

I see the Brahman bull turning. His eyes wandering until they lock the tree line some 50m to my left, seeming to call the bull to break for his freedom. I roll my hands as my body twists to the left, feeling the powerful stock horse move beneath me as if an extension of my body. We start the sprint at the same time, horseman vs Brahman in the blistering dry heat of the dusty red outback station. Lung fulls of hot air and red dust kicked up from the gallop and movement of the huge mob of beefies had become familiar to me after a couple of months here; but time in the saddle and acclimatization still doesn't make this any easier.

My muscles ache, my throats dry but my mind feels free, totally in the moment as hove's thunder beneath me as we take just seconds to block what once seemed an exit for the bull.

Smiling; I pull my boy up and turn my body towards the bull and push him back towards the mob ahead. As he saunters away, I gather my thoughts and dust both my shirt and jeans off. As I trot back to the rear of the mob, I look towards the other stockmen, none who lifted so much as finger to help. A flick on the brim of a stockman's Akubra hat, a smile and a knowing nod said more to me than words ever could. You did good kid.

There's no fences here, and the nearest rider is rarely within 100 meters of me in endless open rust stained grassland, dotted with stands of eucalyptus and patchy low scrub. The Australian cattle stations put my understanding of space into almost a new 4th dimension, in comparison to German. I had only seen the romantic pictures of thousands of cattle being moved on TV, in books and in my dreams growing up. Yet now, here I am: moving 3000 head of beef with 10 stockmen on horseback flanked by cattle dogs. Fortunately, there are a few dogs that are subpar at best, so even my newly acquired partner Jill fitted in, with her enthusiasm far outweighing her ability.

My attention draws back to the cattle, looking up and down the lines of the mob. Our big trip from one side of the property to the other had begun weeks ago; giving the grass a chance to grow in our wake, and the beefies a chance to graze along the way to draft them, brand the calves, and if necessary, cut the horns.

As my horse starts to trot under me without a cue from me, my first instinct is to pull him up; but I quickly realize he's still the better worker of the two of us. The key is to be one step ahead of the stock, so when they try to find a way out, by simply crossing towards it; we block it. My hand wanders to the neck of my horse, stroking him as I mutter "what a good boy" to him softly. It was a lesson I learned that in years to come I would hear Pete say again and again... listen to your horse. I did, and as we continue working our way up and down the line of beefies, I kept listening to my horse as much as my own heart, knowing one mistake could set the whole herd into chaos.

The work there was dirty and honest, with nothing ever working perfectly or to plan. A total contrast from where I came from: a horse world full of money, competition, and jealousy. The first riding complex where I received riding lessons was exactly that, and I was so thankful for "Galoppers-ranch", a stable I found one year later. People there knew each other, talked and helped each other out, and it felt more like a big family than anything else. Western riding developed according to the needs of cowboys, who worked cattle from horseback, and that fascinated me right from the start.

So, I trained for chasing livestock; dreaming about open spaces while cadged in a German arena, constantly improving my technic with horses that have never even seen a cow. I learned how and when to use my legs, hands and weight as well as how to sit straight. Soon I was able to catch and saddle a horse, leading, brushing and picking up hoofs became a routine.

My favorite days were spent riding in the woods around our stable. At first, I would refuse riding anything but the slowest and oldest horse in the stable, all the time within the security of the round pen left me a bit scared. Everything in the real world scared me at that point in my life though, and looking back I feel I must have been sending such terrible messages to that poor old horse.

My trip to Australia finally gave me the chance to put my knowledge into practice, and fortunately my confidence had grown a lot since those first rides. What I expected was so far removed from what I found in Australia; the horses were held in massive paddocks, mustered with a bike into the yards when and if needed. They were still so wild and free; unlike anything I had ever seen in Germany.

My precise commands seemed at times to confuse the horses, and all the hours in the arena, the lessons seemed to count for nothing those first few days mustering. Don't get me wrong, I was very grateful to understand the fundamentals of riding I'd developed over years of hard work back in Germany. But the horses I was riding here thought for themselves, and I often had to just trust in my horse and go with it. Riding became more than a series of commands; it became real teamwork, and it was here when I first felt at one with a horse. It wasn't so much I had to forget everything I knew, but I had to become flexible and adapt to riding with my heart, and not just my head.

Riding next to the stockmen and women out in the bush was simply out of this world; first they really intimidated me, and I only really started to grow when I opened up. I was in a situation where it was sink or swim; rather than being timid, I began to really open myself up and learned to adapt. Again, when I arrived in Central Asia, everything I thought I knew about horses seemed to have no place here. People here were nothing short of glued in their saddles "which were awful at best", with their hands going everywhere, and even the best horses had no manners at all. But man, they could ride!

And ride in such a way it left me once again questioning everything I thought I knew about riding and horses in general. The more time I spent around these wild people, the more I respected them. They didn't have the luxury of a round pen or arena to train in, and there is no such thing as "riding lessons"; it's more a get in the saddle and go from age 2 kind of thing. The result was a connection with horses as natural as walking is for us. The biggest thing though, is that they are fearless...and then I met Pete.

As we cantered side by side across the steppe, we had our first of many chats about horsemanship, and Pete found words that left me speechless and confused...

"You need three things to ride a horse well. You need your head, your heart, and balls."

Kupkari.

The reasons our experience on the steppe have taught us so much can be summed up with one example: Kupkari. Watching and playing Kupkari was an experience I still pinch myself to believe was real. Thousands of years old, it's a game in which both man and horse are often killed as at time thousands of stallions take to the field and fight to drag a 70-80kg dead sheep up and down a field at the gallop before dumping it within a small circle.

There are no safety barriers, and the crowd mingle with the often-galloping wall of horses, and at best the only protection worn by a rider is a soviet era tank crew padded hat. It is a display of pure horsemanship, fueled by vodka and the excited cheers of a crowd screaming in a language I don't understand a word of. Whips fly as a mounted melee breaks out in an attempt to drag the dead sheep up, before breaking free at the gallop, often dragged by a team mate hanging from their horse's reins. Games can last for three hours, and many of the horses have galloped often 40 km to get to the game, and will gallop home again after!

It's like stepping back in time, standing there eating a lamb kebab with a glass of vodka in your hand marveling at the stallions mooching through the crowd when suddenly you become part of a running screaming mass as a wall of stallions lead by a man, dragging a sheep, holding a whip between his teeth bares down on you!

You drop kebab everywhere in a sprint towards a soviet era hay wagon used to truck dozens of these amazing Karabair stallions here laughing as you spill vodka all over your shirt. It's just outrageous; totally and utterly mad. And that's what's so beautiful about it. It's like nothing in the west, and the comments we received after posting videos and images online were often very harsh.

People called it barbaric, cruel and disgusting. But it's not; its anything but. These are testosterone filled stallions that live for this. They are stabled, fed better food than the owner's families, rugged each night and are cared for and revered as the stunning creatures they are. The connection between rider and Stallion is something to just marvel at in stunned, breathless silence.

Kupkari riders view your castrated pleasure pony living in a 2-acer paddock made to trot in circles in an arena as cruel, and more and more I am inclined to agree with this. The horses on the steppe live their lives as horses, and to ride them you must be a horseman, not a rider. You must understand much more than how to sit straight in the saddle with your heels down and hold your reins properly, and it was all this madness on the steppe where we developed these skillsets. Skillsets forgotten or banned in the west as ever growing crowds protest at rodeos, and those utterly disconnected from horses, livestock, and the land dictate what is and isn't considered acceptable treatment of horses.

Kupkari could never be played in the west: a family would be mown down and that would be the end of it. That, and someone would be upset by the use of the dead sheep, so that would need to be changed to a pink bag, and there would of course need to be safety barriers installed, and safety gear worn by all riders and rules and regulations. That PURE horsemanship, and pedigree dating back thousands of years would be at best diluted, if not lost completely... and at far too many riding schools back home, safety has cost us connection.

Luisa and I both played Kupkari, and during training I managed a short canter holding the sheep. It was utterly humbling, realizing if I had trained for my entire life up until this very moment; I still wouldn't be half the horseman many of these riders were.

That famous scene in Rambo three when he plays... utter crap. Theres no way Stallone in his prime could so much as touch the sheep playing Uzbeks, let alone win.

But both Luisa and I took the field, and Lu was the first woman to ever take the field in Jizzakh Uzbekistan on horseback. We had earned their respect, and by god had they earned ours! But Kupkari: its horses, its spirit and its competitors taught me so much about what's required to be a good horseman in contrast to the horse world of the west.

Kuphari horses are all stallions, and must be 8 years old. As a result, the horses are a reflection of their owners as much as their own character, and their bond built over many years of hard work together shows on the field.

The Kupkari riders and their horses have worked dam hard to become masters of their trade, and there just aren't any short cuts or ways to buy your Kupkari master without 8 years of work leading up to his debut. A fact that is lost on many a pleasure rider looking to purchase a "school master" ready trained and perfect for their needs.

Luisa and I both agree many riders are very hard on their horse's mouths, and whips are most certainly used extensively. We are both very soft handed riders who will only ever whip our horses with the end of our reins in a dangerous situation or one where a lesson MUST NOT be lost to a horse. If they get away with it once, they will try it again and we do not let a horse ever cross a boundary we set for it. But none of how we ride is relevant on a monster Kupkari stallion at the full gallop. I managed to hit 68kph on my stallion and believe me it took some stopping!

When riding these Kupkari horses, in particular Sabir's beloved horse "Boing" he warned Luisa not to go past the trot as she would never be able to pull him up. In a mash of colliding horses and riders, flailing whips and violence this is just how it need to be done I guess, but to see the softness Sabir shows Boing after the game is just heart-warming. Both Sabir and Boing LOVE playing Kupkari, and no matter how you wish to view it, I do not consider his treatment of a horse that comes to its name, lifts its feet and shows Sabir nothing but love... cruel.

Love me for who I am, not who you want me to be.

The way Bill the Bastard loved Pete left me jealous.

The way they looked out for each other when things went wrong, and on the other hand their constant mucking around, joking, giving each other a playful hard time underlined the love they had each other. The bond of friendship had become so much more, and I have no doubt that if Pete were born in the late 1890's, he would have given a horse just like Bill the last of his water from his slouch hat before charging the Turkish lines together. Pete struggles to talk about Bill to this day after their separation, and the crazy thing is that straight after Bill, he connected with another stallion.

Beersheba was Pete's project horse in Uzbekistan, and there's no denying he bit off more than he could chew given the severe limitations of our location and time to train the young horse. Beersheba nearly killed him, but there is no denying he loved Pete in such a way it again left me speechless. I remembered the first time we looked at him; a stunning horse, full of life but obviously not handled, and when he had been handled it had been very aggressively.

Every day my dislike for that little horse grew; conversely, Pete's confidence with Beersheba grew and so did his happiness. I saw all that was wrong with Beersheba; Pete seemed to look right through the mistreated and literally dangerous horse, who was so hateful towards people. Maybe not hateful: scared is probably a better word to use, but although nothing like Bill, Beersheba and Pete had something very special and given time and somewhere to actually work with him... Beersheba would have most likely become a "forever horse" just like Bill.

And then there was Gorda, Pete's favorite horse at Lost Ridge, John's ranch where we stayed in Georgia. Even though the two didn't spent much time together, the moments they shared... even a blind man could have seen Gorda and Pete were bonding for life. A cheeky loving little stallion only just trained to saddle, the two of them played with one another like a kid does with a new puppy. Pete would stick the hose in his mouth while washing him down, laughing as Gorda drunk from the hose and spat water everywhere.

But Gorda preferred beer to water; and as soon as Pete cracked a can Gorda, would run over for a drink. Seeing them together was heart-warming, as Pete is all too often in pain or suffering at the hands of unseen demons I can't ever understand, but when Gorda is trying to steal Pete's beer and the two of them are just playing, laughing and having fun... I know Pete is himself.

There's no denying a horse can heal a broken heart like nothing else can.

Georgia offered me an opportunity I'd only dreamed of: breaking horses. It was January, and snow was still falling, blocking access to the places we had hoped to ride, so as mentioned we took the opportunity to stay a while and work. The frozen landscape, the soviet buildings and wild horses roaming the mountain foot hills framed the upcoming experience in the most unique way. While Pete formed another almost instant bond with a cranky filly, I worked with my 2-year-old fluffy little colt, and stunning stallion.

My boys were trusting right from the start with me, and he only had positive experiences: everything they weren't sure about, we walked through together with Pete watching on, encouraging me and most importantly reminding me I could do this! Pete and I had lived and ridden together for six months now, and he was right: I could do this!

It's sad about Alarmi, he was my absolute dream horse. I was so fascinated by his strength, stubbornness, courage and fighting nature; all the things I was lacking, and it made me even more attracted to him; a journey needs to be hard right? You need to overcome obstacles together, otherwise it would we fake; right? I knew I was going to be the first person who rode this stunning horse.

While Kashi loved me more and more each day, trusting me; throwing himself out there with all of his heart, I was almost blind to see that I had connected with him in the same way Pete had so often bonded with horses. And that's where I went wrong over and over again; I was stuck in my head! I thought I had to prove something to others, but even more so to myself. I thought there was this one perfect horse for me, just like Jill is that one perfect dog for me. Even though Jill was nothing like what I had wanted at the time, she turned out to be perfect; Kashi, just like Jill, was everything I needed without even realizing it.

Isn't that what we all are looking for, that super special connection with our horse? All those horse movies such as Seabiscuit, Ride like a Girl, Flicker, Secritariety and dare I say or War Horse and The Light Horsemen "because they make Pete cry" leave all of us wanting, craving and wishing for this once in a lifetime connection with a horse.

I was too busy in my own head to even give Kashi a real chance, and Pete was once more crawling drunk under that same filly that 3 weeks ago looked like it wanted to kill us both. My eyes and dreams were with that beautiful chestnut stallion Alarmi, and in many regards it took his passing for me to see what I had in front of me all along. Kashlama, a true heart horse that is now being gently brought into work as a trekking horse for tourist in years to come. He is soft, kind and loving; and it was my distinct honor to teach both him and Alarmi to be ridden with gentle hands, and soft commands.

Both Pete and I have made a lot of mistakes along the way, but these successes allowed me to finally feel as though I could ride any horse with my head, heart and balls.

Pete will never impress a classically trained rider with anything he does, but he's brilliant with every horse he touches. It's like he looks through all the negatives right through to the horse's heart, and there are most defiantly horse he takes one look at and says in his typical Aussie slang

"Yeah nah mate.'

But what I like most is that rather than trying to bend the horse to fit him, his wishes and picture of how the horse should be; he opens up and encourages the unique character of every single horse he works with.

He gives them the lead they need, and the love and confidence to grow.

He gives them the chance to shine, just how he does with me and Jill.

The ramblings of drunken fool, who most likely shouldn't be giving advice.

My ex-partner was what you would call a nervous rider. Nervous riders make nervous horses, and a lot of instructors make nervous riders. Constantly worrying about where your hands are, heels down, one two one two one two, straight back, etc etc. For me this is not the way to learn how to ride, but for many it works.

Although I didn't think this when I first started riding; I think a rider should start on the ground with a lead rope, doing many hours of ground work before ever getting on a horse. Why? Because learning to connect with and trust a horse is vital, and learning just how easily a horse will move for you, stop for you and respect you is much easier as you look into each other's eyes and your horse forms an image of who they will be working with. It will give you both the heart and the balls to ride your horse confidently, before ever trying to fill your head with technical knowledge that too this day I still don't have any clue about.

Making mistakes is better than faking perfection, and I can't help but feel the way a lot of rider's dress, act and hand out second-hand knowledge is all too often an attempt at this.

The differences between the way horses are ridden on the steppe and in the west are endless. On my return to "civilized" society I am awaiting endless criticism for my attitude towards horsemanship, forged among wild men, in wild places with even wilder horses. I may be completely wrong about everything I say here; this may bare no correlation with conventional wisdom towards horsemanship, but it's the lessons I speak of here that have personally helped me in my journey.

I cannot stress enough, I have had no formal instruction as either a rider or horseman; I have learned almost exclusively on wild horses for the most part barely trained if trained at all, and there is one lesson I have learned from this above all others. If your horse is doing something you don't like, it's because you're doing something it doesn't like. You need to change, in order for the horse's behavior to change.

Horses are not like dogs, who's loyalty can be purchased with a scratch behind the ears, the belly and of course food. This isn't enough for a horse to become a lifelong partner, trust must be earned gradually, consistently and as a result both Luisa and I agree a 5% gain with your horse each day is our goal.

The difference between raw beginners and those that are capable, competent, and handy with horses is often the number of mistakes that a person has made and learned from, rather than the lessons others have taught them. The more you grow as a horseman, the mistakes you make become fewer and the consequences are less damaging both physically on your body, and more importantly on setting the horse backwards in its development and growth as your partner. Lu and I have made plenty of mistakes, and now we have reached a level where it's rare for us to have major setbacks with our bonding with new horses.

As you bond with, or train a new horse, there is inevitably a point where a horse changes their view of a person, but not necessarily people in general. Luisa and I have had both gradually transition from being viewed with distrust, to someone whose intentions are worthy of consideration, as well as instant bonds with horses. The point where it changes is different for each relationship, each horse, and of course horseman.

We have both struggled to get along with a horse at times, and the point where you transition from butting heads to smooth partnership generally has a sort of clear tipping point. On the other side of this is where the partnership really begins, and the real gains in both training, friendship, and even love begin. In this place, the reasons for becoming a better horseman become clear, as it's here you share time and adventures with a partner like no other friend you will ever have.

Every relationship has its boundaries, and your journey with a horse never ends; and it's the journey itself where both Luisa and I find the greatest happiness. Your journey with a new horse begins the second you meet, look at one and other, touch one another and most importantly set boundaries for one another. For me, the way we touch a horse is vital, and it was my Mongolian mentors that taught me how to calm a horse and touch it correctly.

Mongols whistle softly to their horses often; with a short high tone then long low tone following. The Tush horsemen make a soft brbrbrbrbrbrbrbr noise; and I've always used a short "Hey" followed by several soft "hay hay hay hay hay hay heys". As you reach too your horse, allow him to approach wherever possible, although with a new, unbroken, or green horse you may need to use its head collar to draw it closer. The Mongols will always offer the horse a chance to touch you first, and if it does, this is a good start. From here gently whistle, brbrbrbr or use whatever soft calming noise you feel most comfortable with as you run your hand softly down its neck.

There are of course a ton of subtle cues a horse will give you at this time, and learning to read them is something that doesn't take a lot of time at all. There are a few key things a horse will do when you approach that give away how it feels towards you, and I feel it is very rare for a horse to attack you without ample warning. If it does attack without warning, you've most likely scared it or done something you shouldn't have, and most of these things can be avoided easily.

Ears are of course a dead giveaway, but one thing that I feel is really important to mention here is how a horse stands before you get on it. If its leg on your side is relaxed and it's attempting to stare you down... often with its ears pinned back, you're in for a fun ride. But if its off side leg is relaxed, it's a sign the horse is far more accepting and willing to cooperate with you. If I see signs that a horse isn't ready to be mounted, I will where ever possible spend some time doing ground work with the horse before mounting it. It's far safer to make the horse walk trot and canter several laps of you, and wait for those subtle hints it's going to try something to disappear than force a connection prematurely.

When a foal is born, it falls free to the soft earth beneath it, where its doting mother will immediately begin nuzzling it and licking it clean. The Mongols have made it clear to me, this first memory is precious; you should remind your horse of its mother and this moment. Your hand should run down your partners neck like its mother's tongue liking it clean as you make soothing noises. But. You must be like a boss mare, not a mate yet, and pats and love come with boundaries.

My boy Digger knows I'm not a horse; he's not dumb, and he doesn't need to look to me as his leader, boss or anything more than something that makes his life easier. He has his herd; his boss mare and he is, after all, a horse. He doesn't have to cooperate with me at all, and if he chooses to it is because I have shown him that the path in which I am steering him, is the one that requires him to do the least for the greatest reward. He knows about the bears and wolves, and he knows I have a gun, no fear, and treats, pats, a brush and cuddles. I remind him of his mother and set boundaries just like a boss mare.

We have come to an "understanding" with each other, and I have in no way broken him.

He moves like water, following the path of least resistance and he will always move away from pressure; be it the bit in his mouth, bees bursting from an unseen hive he stepped on or a gentle heel in the ribs. Digger was the first horse to ever sort of "talk to me", and nothing annoys him more than being tied up, as it teaches him nothing. He told me he gets bored; and that he will stay with me as long as I didn't push him away from his friends or everything that makes him a horse. If he doesn't want to come over, it's because it's more effort too than not, so I reward him when he comes.

He likes to be touched too; he tells me where and when and loves to listen to stories as to be honest, even though he knows I'm not a horse, he enjoys companionship as much as I do. He likes calm voices, and understands tone and intent very well, but it's more than just that: Digger told me he can feel what you feel, and an uneasy rider makes an uneasy horse. So, drink your vodka Pete and relax.

Digger tells me he thinks about me, sees me as in partnership with him and trusts me as much as I trust him. He still comes to me in my dreams, nudging me with his nose as I snooze in the long summer grass of the Mongolian North. I can hear his soft breaths and feel his warmth on my face. Sometimes; when he wakes me up, I can almost smell him.

I've read a few books on long rides, but never read a book on horsemanship as such, or watched YouTube videos hosted by a guy in a cowboy hat with a microphone talking in an arena. I went out there and I learned how to touch and handle a horse first hand, and I was only able to do this because mentally there's something wrong with me.

I have PTSD. I've battled with it most of my adult life after a ton of traumatic events I won't get into, but it imbued me with a strange sort of calmness in the face of things that would terrify others. I suck at interpersonal relationships, am anxious in cities, but... I just kind of go cold in the face of death or danger. I can think, my brain becomes clearer almost, and although I will often fall to bits after the event has passed, at the time I have no real fear. It's kept me alive many times over, and it is this lack of fear that has helped me become the horseman I am today as I can keep a panicking horse calm. As any horseman will tell you, only one of you are allowed to panic at any time, and it's never your turn to panic on a horse. Ever.

Managing your fear around and on horses is important for a ton of reasons; but the biggest is this. A horse is a prey animal; wolves eat them, bears do too, and leopards, hyena and lions all hunt horses the world over. It's programmed into them, and if they get a hint of fear from you, they will begin to look for what you're afraid of. They will be on edge, and that scary plastic bag; will become a charging bear. But if you're not afraid, if you're the boss mare, then when a horse spooks you calm it with your confidence, a soft whistle, a brbrbrbrbrb or hey hay hay hay as you touch its neck and reassure it.

Do it often enough, and your horse will desensitize to dam near anything and stay as calm as you do.

BUT: if you whack your horse, "and of course there is a time for discipline and reinforcement of boundaries" your horse may come to fear you and behave out of fear of discipline. This is not idea; because the day your horses sees something it fears more than you, you'll have no control at all. Fear flows from you, as it flows from the horse and the more time, I spend around them I believe more and more in some sort of interconnecting energy between all things. In fact, yesterday Luisa and I worked with a wild 5-year-old painted stallion whose intensity was such we could literally "feel" him from a distance as though there was some sort of electricity in the air.

But there are times when I will hit a horse with my reins, and it took me a very long time to get used to it as I was far too soft when I began riding. The Mongols taught me how to physically dominate a horse, and it was without a doubt one of the most terrifying experiences of my life.

It was the one and only time I rode a completely unbroken, wild horse that had never even had a bit in its mouth before. A gorgeous plain grey pony with a dark mane; it had been caught with a lasso on the end of a long stick at the gallop.

As I galloped alongside, pushing the mob in range of the other horsemen, a rope was slipped over its head and the Mongol horseman dropped from his horse. In a split second, he ripped the wild horses head to the side, and ran around it as fast as he could; dragging it off its feet with a savage yank on the rope. In seconds he was on it, tangling it in the rope as he yelled at the other horseman. The horse he'd just sprung from just stood there, ideally watching the whole procedure as a head collar and bridle were thrown over the wildly flailing horse.

I dismounted, taking up the lead rope of my friend's idle horse as the grey pony found its feet and began to viciously circle the Mongol who had it by the lead rope as he tried to throw the reins over its head. In a split second he grabbed a handful of mane, and seamlessly mounted the bucking and pig rooting wild horse. Rather than ride the buck out he belted the horse as hard as he could across its rump with the flailing lead rope yelling "CHEW CHEW" to cheers and laughter from the rest of us.

As he began galloping the mad little horse flat out across the empty steppe, I was laughing in a sort of dumbfounded shock as to what I had just witnessed. A large bowl of mare's milk vodka was handed to me by a grinning Mongol nodding in approval, and I gratefully accepted before handing it to the next smiling, laughing horseman. I was astounded by the Mongolian rider's ability I'd just witnessed, particularly given he was wearing a full-length Mongolian "Dell", a traditional garment that is both bulky and heavy.

But I was even more shocked as before I'd even finished choking on my vodka, the grey pony was trotting back towards me with the rider casually laughing and chatting away to the other Mongolians. The little grey did little more than shake its head and object slightly as he dismounted and began stroking and calming the horse. I approached to congratulate him as much on not having died, as having ridden the horse when the strangest thing happened... he just handed me the reins and said

"Hazaar, chew chew!"

Hazar is the name I am known by in Mongolia; and in that moment I knew despite being eye to eye with a horse that clearly wanted to kill me... I had to ride it.

I grabbed the mane, threw myself over the horse and didn't even need to say chew before it bolted. I'm not sure if I even had the reins, and it was all just a blur as the powerful little beast charged off towards the horizon with my only thought being to keep my balance bare back at the gallop, and keeping the horse bolting, rather than bucking.

Eventually he started puffing; I found my reins and slowly started giving a little mouth pressure and leg to turn the galloping little lunatic in a long slow circle. Heading back towards the other horseman I was pretty happy with myself for having even stayed on, and now to my utter horror, noticed the saddle had been taken off the horse I had been riding that morning.

Ahh... OK. So I'm riding this thing now.

I trotted a few tight circles, halted my horse and backed it up before dismounting and stroking its neck. Touching a Mongol horse's face isn't a good idea unless you know the horse well, and going down its right side is a great way to get killed. Again, I was handed a bowl of vodka to cheers of

"HAZAAR! HAZAAR CHEW CHEW!"

As I choked down more mare's milk vodka, my hands shook a little, and I noticed substantial rope burn to which I had no idea of its causes as my Mongol friends struggled to saddle the little grey monster. They were whistling, and doing all they could to calm the little beast that didn't have even a hint of terror in its eyes; on the contrary, it only had a look of stoic defiance.

Once the saddle was secured, I was again handed a bowl of mare's milk vodka on which I again badly choked; coughing and spluttering as I approached the mad little grey. Three people held it as I gingerly made my way into the saddle as he fizzed under me. And; that was it... the Mongols threw themselves into their saddles, and we were off; cantering in a mob across the steppe. What on earth was I thinking... and believe it or not, a Mongol horse has never binned me hard and that little grey didn't so much a put a foot wrong for me!

So, what's the point of that story? I honestly don't know. And I doubt there is any aspect of it that can be used in any way of value for the average rider reading this book.

But horses know when they're not going to win, and they know when it's easier to just comply and "come to an understanding" with its rider. The Mongols are masters of this, as many have thousands of horses in their valleys and you need to be able to just grab one and go when needed. Of course, they know which horses will become food for the long winters, which will be raced, gelded, ridden or lost to the wilds. But this style of horsemanship isn't one involving clean movements in the arena, jumps or any fancy movements by the horses themselves.

It's just wild and pure; its where I learned to ride, and ride anything for that matter.

Harsh realities.

Firstly, I would like to say without a hint of hesitation, Luisa and I love wild horses that know as little about people as possible. If you meet them when they are young, they are full of fire and they have spent their lives being a horse. No fences, just the rules of the mob and they know how to both look after themselves and have established boundaries within their group. They are simple, beautiful creatures to work with once they have learnt to accept a relationship with man, but will always retain that heart of a horse born wild and free provided you nurture it, and don't crush their spirt.

Pippa, my wild Georgian mare that took us a full 3 hours to finally run into the stables, and a month to train to saddle looked at me like a hungry velociraptor when stabled for the first time. She absolutely wanted to kill me, but after three days she was halter broken and loved being groomed and touched by me. She was soon giving me her feet and was easy to sack, happily letting me drape rugs over her body and head. She spent a few weeks listening to Aussie hip hop in the barn walking about with a saddle on, and was soon becoming desensitized to the ways of these strange two-legged things that sat on her friend's backs.

Ground work followed, lots and lots of ground work. We had no round pen, just bare snow-covered winter fields surrounding the crumbling stable block, so those first rides were gentle bare back experiences on a lead rope held by Luisa, but there were no real problems and each day we made our 5% gain with the stubborn little mare. Then, she breathed a deep sigh one morning as I mounted her; and that was it. She had come to an understanding with me. Better horsemen than us can do this in days, but we feel anyone can do it themselves in a month or so if they have the basics sorted.

That little mare and so many other horses along my journey have each taught me about riding, not riding one well trained horse to look pretty in an arena; but riding anything from A to B in such a way both rider and horse feel safe and respect one another. Make no mistake, we consider dressage an art form that requires the same commitment as Kupkari, with grace and soft subtle movements in total contrast to the brutal nature of Kupkari.

How and what discipline you ride is up to you completely, and no matter what you're doing I will always marvel at a horse and rider working beautifully in a seamless partnership. To be honest, barrel racing and cutting are just as exciting to watch, and so is high level dressage and jumping. The point I think we both really want to make is all about a rider's connection with a horse, no matter what you're doing; it's the connection we respect, and there are many ways to achieve this.

We can only speak of our experiences, and offer this advice as just one of the multitude of ways to connect with horses. I've learned that because I am utterly fearless, 65kg, and very powerful for my size, I have the mental and physical attributes to ride well. Luisa is also 65kg and although a woman, she is in no way delicate or fragile, but it is fair to say she takes fewer risks than I do and relies more on technique than strength. We ride completely differently, and can get the same results from the same horses with totally different cues.

There isn't just one way of doing things, and finding what suits you and your style is a journey you must take on your own and you must be realistic in it. I'm not saying "you can't or you shouldn't", in fact quite the opposite. I am saying you can, and you should; but ride within both your limits and the limits of your horse. Having the technical knowledge will only get you so far, and for me personally I have had a long journey getting this up to par with Luisa. But I've always had heart, and I've always had the balls to step out of my comfort zones...

If you can both be empathetic and kind, yet fearless and dominate when needed, you can establish an equitable relationship with a horse. But never forget: horses don't make mistakes, we just don't speak to them properly, and when something goes wrong, and it will... it isn't the horse's fault, it's yours. If you are willing to take a deep breath, accept each and every day that what you love may kill you; if you feel you will never stop learning, improving yourself physically, mentally and emotionally. Then you have everything you need to become a great horseman, regardless of how good your gear or horse is. Knowledge grows over time, and neither of us will ever stop listening or learning. We love what we do, and I are prepared to fall doing it.

Chapter 7.
Into the wild.

"We all strive for safety, prosperity, comfort, long life, and dullness."
Aldo Leopold.

I've spent at least half of the last 20 years in hostile environments of one sort or another, and there's something pure, enjoyable and in a strange way, spiritual, about succeeding when things really suck. I want life to be hard. I want to carry beautiful memories of adventures made even more beautiful by the pain and suffering endured to experience things others only dream of from the safety of home.

Coming into this journey we had a very big advantage, and that is the years I'd spent in the military, as an adventurer, and most importantly, as a hunter. Every night I spent in the field, I learned something new. And when I was coldest, soaking wet and cursing some utter piece of crap that had let me down... I had trips I look back on as my greatest successes. This time in the field taught me both to mentally prepare for and endure hardship, but also to equip myself with the right gear for the job. Luckily for us, this wealth of knowledge acquired over 20+ years translated to a very sound base of tools as useful on horseback, as they were stalking deer in the land of the long white cloud, or wolves in the land of blue sky.

Luisa and I are both good horsemen, listening and learning daily and constantly evolving the way we ride and live our lives on horseback. We have had to evolve the gear we use on our horses just as much as the way we've had to evolve the way we ride and equip ourselves for daily survival. We know more than enough to know we don't know everything, and that there's a journey every horseman needs to take in their own way. Good horsemen never stop learning, and we have loved every single minute learning these lessons. So, from here let's offer some advice, the advice on horse gear the two of us felt we needed right at the beginning of our journey together.

I've been very fortunate in the opportunities that have presented themselves to me over the years, and it never ceases to amaze me just how differently people hunt, and endure the elements around the globe. I've hunted alongside professional hunters and other guides with all the bells and whistles, barefoot natives in nothing but shorts that spear mackerel with a sharpened bicycle spoke hurled into the shallow clear waters of the South Pacific islands, and nomadic herdsmen in traditional hand sewn garments made from skins that hunt wolves with rifles almost a century old.

The goal is always to harvest game and brave the elements, but the ways in which we go about doing it, and more importantly the amount we have financially invested in it varies incredibly. Far too often the hunting shows have sponsors that promote their gear at every single opportunity from the low angle shot of boots walking in, to the backcountry cuisine in the heavily branded tent which is such a stark contrast to the way I've seen natives making do with all they have available. I think the perfect combination of gear is a mix of old and new, and you defiantly get what you pay for. But paying for it isn't always best, sometimes you need to craft your own, and know how to repair your own gear to make it in the wilderness.

Fiordland, on the south west coast of New Zealand is unlike anywhere else I have ever hunted, and it is without doubt the most valuable of all training grounds in which I have tested and destroyed gear. The bush reminds me somewhat of the South Pacific jungles during the wet season, in which I've spent much of my adult life. Everything in the jungle is wet, and Fiordland is exactly the same; only Fiordland is cold when it isn't freezing, and the sun often fails to clear the ridges that plunge near vertical into the fiords.

The faded drab olive greens of the beech trees, the soft dark leaf litter of the forest floor and the rocky brown snow grass tops are all in stark contrast to the emerald valleys of the South Pacific. Fiordland resembles "Pandora", the mythical planet from the movie Avatar. The wildlife here is also unique, most of which has no idea what on earth you are, with wekas and keas "flightless birds" showing you little interest as they go about their business.

Despite all this beauty, it really sucks being in Fiordland for any length of time unless you are properly kitted out, and physically, emotionally and, most importantly, mentally prepared for your time there.

Mentally Fiordland poses challenges that seem to have been glossed over by magazines for a long time. Every morning you need to embrace the suck of the place. Slipping into yesterday's wet clothing after a cosy night in a warm sleeping bag listening to the rain hammer the tent is hard. Often too hard. From there you get cold and wetter, but provided you keep moving, it is relatively comfortable bush to hunt, though the push to the tops above can be extremely challenging physically. Once you make the tops, you look at your watch and realize you've burned 60% of the daylight, and a little voice in your brain says

"Hurry back before dark. If we spend the night outside, you'll die."

For many hunters these fears, apprehensions and perfectly reasonable thoughts keep people close to camp. No matter how inviting it may look on Google Earth I assure you the second you set foot in Fiordland, all bets are off.

Spending a year of my life in the Central Plateau of New Zealand's North Island training with the military prepared me very well for my life of adventure in harsh environments too. It's not completely unlike the steppe, and spending up to a full month at a time living out of a pack, walking all day, digging a fox hole each night and constant sleep deprivation help you realize a lot about yourself. You don't know how far you can really push yourself until you've had to keep pushing past anything you thought possible in order to survive.

Strangely, after days when you've gone 56 hours without food or sleep covering over 100km on foot with 40 kg of kit and a rifle... life; just kind of seems easier. Deployment taught me a great deal about myself too, and I'm a big believer that you don't need to be a hunter, outdoorsman, former solider to do well in the outdoors. You just need the right attitude and gear that will do the job, so let's talk about gear.

Before I talk through a full suggested gear list for a horseback adventure, I am going to state clearly that your chosen level of comfort may differ greatly from ours. For us, me in particular, less is most definitely more. When we started we had all sorts of crap. Crap we never needed and gradually whittled away as we were constantly forced to adapt.

Losing our pack horse was the best thing that could have happened to us, and the way we evolved to be a pair of riders on a single horse each has made our travel faster, easier and far more enjoyable as a result. We firmly believe that you can easily survive comfortably with limited gear well balanced on a single horse for long periods of time with the following gear.

Luisa and I have no sponsors. We do no paid promotions, this isn't a business for us, its a lifestyle and as a result we're pretty much broke. Every bit of gear we recommend is because for us, it represents quality and functionality we put our trust, and literally our lives in. In the Acknowledgments section at the end of the book, I have provided contact details and links to these amazing products we live in our use daily.

Full gear list breakdown.

Individual Clothing list.

Merino wool socks x 4 pairs
Merino wool long johns
Merino wool long sleeve tops x 2

Merino wool beanie
Merino wool buff
Shamar scarf

Wet weather jacket

Wet weather pants

Light micro-fleece long pants
Light puffer jacket
Durable long sleeve shirt
Durable long pants

Micro-fleece T shirt
Light wind proof jacket
Cotton T shirt
Baseball cap / or wide brim hat

Hiking boots

Riding chaps

Luisa carries 3 sports bras and 5 pairs of underwear in addition to this list. If you are of a C cup size or larger, wearing 2 sports bras while riding is a great advantage and strongly recommended.

Additional winter kit for conditions below freezing, comfortable to as low as -20 Celsius.

Lamb skin gloves
Heavy puffer jacket
Warm sleeping bag liner

Heavy woolen socks
Wool or polar fleece mid layer
Horse blanket

Sleeping and protection from the elements.

Seasonally appropriate compact sleeping bag
Compressible dry bag to store sleeping bag and waterproof roll bags
Suitable tent or light weight tarpaulin

Jet-boil stove with adapter for Asian gas cans
One light weigh cooking pot
Plastic bowl "preferably that folds flat"
light metal utensils "plastic ones always break"
Stainless kebab skewer and Tin foil
Tupperware container with a good screw top lid for leftovers

SHARP long bladed hunting knife
Tomahawk "suitable for use as a hammer"
Folding pocket knife
Repair kit "to be discussed in detail"
Med kit "to be discussed in detail"
Veterinary supplies

Toilet bag "as minimal as possible"
Chap stick, sunscreen and medicated talc
Hanging type fishing scales
Good quality power bank
Solar panel charger
Smart phone with Google maps
Head torch

Optional extras.

Good quality camera	Plastic rain poncho
Lightweight laptop	USB speaker
Bow and arrows / with fishing ability	Telescopic fishing rod

I will discuss the fishing and hunting topic in a later chapter and the option to carry a good camera is completely up to you. We have a reasonable camera, but barely ever even get it out of its waterproof bag. We find our phones do a pretty dam good job of the pictures, and they are always handy. I have a Mac Book air which we have both really enjoyed carrying too, it's allowed us a chance to both work and watch movies together when we stay in guest houses and during bad weather breaks.

Recommended brands and products.

Sue from Swazi has saved my life twice; Sandra has been the difference between life and death too. I know this because every Swazi garment proudly has the name of the person who crafted it written inside it on the tag.

I pay the extra, and I buy Swazi gear, I always will. I have gradually replaced almost all my outdoors gear with Swazi over the last 16 years. The reason for this is all the other gear has worn out, become torn, had zips break or lost its waterproofing. The Swazi stuff hasn't, it fits better, it lasts and if you have an issue with it then send it back and they fix it for you free of charge.

The Thar and Wapiti coats are in my opinion the single most important piece of clothing an adventurer can own. In no less than half the images of Luisa and I, we are wearing a Swazi coat and there's good reason for it. They are brilliant, and when you flip that hood up on the windswept steppe in -20 degrees it's like putting on a space suit. They work. They just work so damn well and I will continue to tell every single person who will listen to get online and buy one.

My micro fleece tops, pants and even my socks are Swazi. As for the socks, I tested to see how many days they take to become unwearable in field conditions. After 27 days on my feet, they were fine, I was fine; they smelt no worse than damp wool. I rate the Swazi merino sock as my single most valuable item of outdoor clothing I own due to their indestructibility and warmth.

Merino is the other secret weapon alongside Swazi when it comes to comfort in the field for me. It is ALWAYS the base layer regardless of where I am for several reasons.

Merino doesn't stink, and scent is something that simply cannot be down played when hunting or more importantly sleeping beside a beautiful woman in a tiny tent. But the magic of merino is seemingly endless; it keeps you warm when it's wet, and you are far less likely to develop skin irritations or sweat rash with a natural fibre against your skin.

The most important thing it does though is keep you warm, very warm in fact and layering of clothing is something few new to living in the outdoors really understand fully. There are a few good tricks to know, the first of these is if you want to stay warm, you never wear anything but merino as a base layer. If you wear a cotton t shirt as a base layer, the effectiveness of everything you put over this is reduced dramatically. If you want to stay warm, merino should be the first thing you put on every time, keep the cotton T shirt for a day you want to stay cool.

We wear Icebreaker merino long johns and long sleeve tops, and sure... they have a few stitches holding them together, but many are over 5 years old! Iv'e always loved them, especially while hunting in New Zealand where I would cover huge distances on foot each day from a base camp deep in the hills. Two long sleeve merino tops screwed into a ball fit in your pocket easily when it heats up, as I prefer to hunt with nothing more than a rifle, knife and a few muesli bars and even drink from the creeks to avoid carrying a water bottle. I'd often bone a deer out completely, tie the sleeves of a merino top to itself, and use it to carry all the meat out as a backpack. These same ice breaker merinos are still going strong even after this endless abuse.

But it's not always cold, and I've spent years sweating in summer heat or slugging through the jungle wearing anything but merino and Swazi fleece. Sitka, another brand you pay more for stands head and shoulders above any other light weight gear I've worn. The clothing is made with an incredibly durable material with 4 way stretch to it and silver fibre wicked through it to reduce scent.

I've put this gear through hell under the same conditions I put an issued NZ army uniform through. In less than 6 week of jungle patrolling, our uniforms were thread bare, torn and literally transparent across the shoulders where our packs rubbed. Under these same conditions I have subjected Sitka gear to a full 2 years of brutal tropical conditions and have only had to repair one tear caused by a solid hit from a machete!

When it comes to your own choice of clothing, fit, style and comfort are all up to you and availability will vary depending on where your based and what your budget is. Less is more as I've said before, and I cannot stress enough though; buy the best gear you can afford as your life may depend on it.

And be very careful who tries to sell you gear; for example, I don't like a certain "entertainer" who hosts a very popular outdoors show. He is just so over the top, that and he stayed in a hotel while filming in New Zealand. His merchandise is rubbish that sells because his name is written on it, and although fine for a weekend trip, I would recommend a far better example of an outdoorsman. The great and powerful Ray Mears quietly picks mushrooms and herbs rather than eating worms and drinking his urine, while wearing the same Swazi coat I've sworn by. Ray Mears is a man I admire, respect and would recommend others look too as I do.

Speaking of others I respect, Bijmin Swart, who is currently riding from one end of New Zealand to the other on wild Kaimanawa horses with his partner swears by rain ponchos over the top of everything. I've thought on this a lot, and believe it will be something I will test out myself this coming winter over my Swazi coat. Having a durable 100% water proof light weight plastic poncho seems a good idea to me especially given my saddle being covered by lamb's wool. I've been told it can be particularly useful as a way to keep your hands dry too, as well as having the ability to run water straight of the rider, his thighs and saddle in a heavy down pour.

I take the advice of other riders very seriously, and have even watched a ton of old westerns in an attempt to figure out new takes on old ideas in our search for a winning formula on long rides.

I wear good European hiking boots and always have over riding boots. There's a few reasons, but the biggest is I lead my horses a lot, particularly in the morning and evening to warm up and cool down the horse prior and post riding. Durable, reliable and very comfortable they are not for everyone, but with the addition of cages to my stirrups I feel safe and comfortable riding all day in these in conjunction with a pair of gators / chaps that mean my leg hairs and skin stays right where it's supposed to. They were cheap too, from a hunting store not and equestrian one.

As with the boots, hunting and survival gear is focused on durability over style, and while endurance racing in Mongolia I saw several peoples riding chaps fail, where my rugged green hunting gators are still going over 2 years on with no visible sign of failure on the horizon.

Cooking is something I've come to love more and more over the years, especially when its game I've killed served with food foraged locally. I'll give a few pointers on just how easy it is to reliably find food later and even a few of my favourite recipes, but for now let's talk about the Jet-boil and why it is a piece of kit I swear by.

Jet-boil items can be purchased online, in any good camping or outdoors store and it is just bloody brilliant. It is incredibly efficient and can heat water faster than anything else on the market and that matters a LOT. If your cold, let's say you've come off your horse during a river crossing and hyperthermia is a risk that's now a real reality... the sooner hot water is inside of you, the sooner you can stabilize your core temperature. It's also using a lot less gas than a conventional stove too, and that means you need to carry less weight in gas cans. Just don't cook food directly in it, only boil water in it and use your pot for cooking stews and the like.

Also be sure to get an adaptor for the point where the gas bottles screw in as Asia uses a completely different type of gas nozzle and the two aren't compatible. I cannot stress this enough... Purchase a jet boil and a nozzle that allows you to uses either type of camp gas.

We keep pots, pans and other cooking items to a real minimum and don't feel the need for a bunch of specialist items. We cook in one pot for most meals, and when we have the chance to enjoy a camp fire at night, we go all out on enjoying ourselves!

Tin foil is pre-cut and folded neatly into squares so it sits nice and flat in our saddlebags and is just a dream for wrapping potatoes and other root vegetables. Fish go beautifully in foil too, as does one of my "made Luisa fall madly in love with me" meatloaves. Wild herbs, a chopped onion, and egg and minced meat slow roasted beside some potatoes and fresh bread with butter... dam. You just can't beat it; the crackle of the fire, the horses grazing nearby as we gaze at the stars drinking whatever local booze we've found in the last few days.

It's pretty dam hard to beat, and if you're lucky enough to have a fresh rabbit, or some meat then a stainless steel kebab skewer, a sprinkle of salt and pepper and hopefully a bottle of red will just breathe the lust of life into your entire being. But that finishing touch once the suns gone down; music, and it's for this reason we carry a UE BOOM2 Bluetooth speaker with us. It's water proof and lasts a long while between charges, and trust me, seeing Luisa dance around the fire after a few vodkas to German songs makes this little speaker worth its weight in gold.

I review a lot of gear on my Facebook page I have for my old hunting business and when the topic of torches having let me down so often came up. The out pouring of hate towards a very popular brand, despite heaps of "sponsored hunters" saying they were great was overwhelming. The verdict is again "spend the money" and get an Olight, Max torch, or Fenix head torch of the highest quality you can afford, rather than a cheaper item masquerading as high end.

Having a nice light weight tomahawk is essential for a couple of reasons. We used ours every day hammering in the spikes we tethered the horses too every night, and it is very handy knocking in a tent peg in frozen ground too. On more than a few occasions in Kazakhstan we were hacking through inches of ice on a frozen puddle, stomping holes for the horses to drink from.

A set of small fishing scales are essential for weighing and balancing gear. I will cover the importance of balance soon, but a few dollars on a set of scales is absolutely invaluable due to the constantly changing weight of your gear as you travel. Food eaten during the day, layers of clothing taken on and off, and the extra weight of wet gear can unbalance your horse, and the few minutes extra spent checking the balance before heading of is time well spent. We rarely have any more than a few hundred grams difference from one side of the saddle to the other, as a result the chances of injuring the horses are greatly reduced.

Repair kit "in detail"

It's not so much that things break, but things aren't working as well as they could be. Our gear has evolved weekly, if not daily as we have found better and more efficient ways of doing things. As a result, I've learned to sew and sew very well for that matter. In any outdoors shop or good equestrian shop you can purchase an item called a sewing awl. It is a sturdy needle on a wooden handle that is very simple to use and can pass thick, waxed nylon thread through heavy canvas, webbing and even leather. It can fix a torn pack, saddle, or sew a strap or clip back on easily. This tool and knowing how to use it well is one of the most vital pieces of advice I can give.

We also carry lighter sewing gear, and I have sewn my clothing back together in more places than I care to think about having worn the same pants for literally over a year. I feel really comfortable doing it and proud of just how long we can make gear last in a disposable world. That and personally, I find something very rewarding about sewing a torn saddle bag by firelight listening to the horses and a fish cooks on a stick and I sip a little whiskey.

Super glue and electrical tape are two other items I find are just magic to work with. Tape can be used on a bunch of stuff for obvious reasons, but on a small tear or hole in a waterproof bag or tent, a liberal application of super glue to the sticky side of the tape will mean that it's never coming of and that leak is a thing of the past. To be safe I put tape on both sides of the hole forming a solid patch that prevents further tearing of the hole and believe me, a ton of our gear has little bits of black electrical tape glued on it!

A good length of para-cord is always handy to have, and a few buttons and perhaps a replacement plastic clip or two isn't a bad idea either. But the big thing... sewing awl with at least 5 spare needles! Another handy trick is if you use clear zip seal bags to keep things dry, a layer of wide electrical tape over them means they last a lot longer and are not prone to getting holes; a cheap easy way to make another long lasting effective dry bag. I particularly use these for documents and the electrical gear as another layer of protection within a standard dry bag.

Veterinary supplies

Many of the items you can carry for a horse are things that can also be used in your own medical emergencies, but there are a few things I carry that are horse specific. There are items that are easy to get overseas, and items that are dam near impossible to find anywhere. You will have no trouble getting needles, penicillin and saline in the event of a wolf attack, large cut or similar injury requiring antibiotics. If you need to sew your horse up, you can easily get Novocain, a good local anesthetic and the gear to stitch your horse up. Sounds crazy... but I never thought I'd be a vet until I was delivering foals and dealing with wolf ravaged horses. I'm not saying expect it, but there's a few more things I wish I had on me.

The top of that list is Bute. Talk to a veterinarian before heading off and do your best to get yourself a large tub of oral Bute and talk about its use. It is amazing stuff and I swear by it for everything from a stone bruised foot to a strained muscle. Be sure to carry a bunch of rolls vet wrap too, you won't find them in Asia sadly.

The other drug you will struggle to find is oral Ivermectin or similar and coming into spring, particularly if the grass is a bit late coming up then worming and treating for ticks is a real boost to your horse. We picked ticks off ours at a rate of as many as 100 at a time and they end up on you too! There is Lyme disease here, but the treatment for it I also carry in my med kit which I will cover in a minute. A few tubes of oral drench are impossible to get here, and the effort we went to getting some sent to us was ridiculous. Be smart, bring them with you instead.

We carried the famous "purple spray" as well as white ointment to begin with, but they were bulky and kind of a hassle for the amount they were seeing use, so we switched back to honey. Honey has been used to treat wounds for thousands of years and it is available everywhere, goes great on bread, in your tea and is the perfect thing to treat a saddle rub or cut with. So for us, it's now Bute, Ivermectin, and rolls of vet wrap we consider the essential must carry kit for a long rider with anything more you may need sourced relatively easily.

Medical kit "in detail"

We take every effort to avoid an injury requiring medical attention, and as a result carry a light kit focused primarily on saving "life and limb". If you get shot, chances are that's it sorry, and when in wild lands full of wild men and beasts... it can happen. But you can die crossing the street because you're on your phone, COVID-19 can get you, and eating cheeseburgers will kill you slowly.

Horses... horses can absolutely kill you and in a hell of a lot of different ways. The more time you spend around them the less likely you are to be injured and I would strongly suggest before embarking on any adventure such as this you do a detailed Saint John's medical course. For example, did you know if you violently cough when having heart attack symptoms, you can possibly prevent it? Knowledge is the best medical kit you can carry and with a few basic items you can save life and limb very easily in the event of a medical emergency.

But let's cover our biggest risks as riders and how to treat them in such a way we are able to get to higher aid. The biggest risk is of course spinal or neck damage, with broken bones right up there too. For these we carry an amazing item called a SAM splint. Its light weight, foam covered malleable aluminum that can be used to make a neck brace, arm split or even leg splint easily and quickly. For serious bleeding we carry gauze to plug holes, Israeli bandages, and the indispensable CAT "combat application tourniquet".

I carry super glue to close wounds, and Betadine for both myself and the horses as well as trusty old vet wrap. I also carry something a little more serious for the "worst case" scenario. I carry chest seals for the event of a punctured lung and a needle to decompress the chest cavity in the event of serious chest and lung trauma caused by a horse rolling on me or impaling me on something. Have I used anything from the med kit other than band aids and Betadine? No, but I'm bloody glad I have it and even more glad I know how and when to use it.

Ciprofloxacin is a very handy drug to carry, as it will kill dam near anything that has you unable to keep food in either end of your body. Having the ability to treat yourself quickly when miles from a doctor, that most likely cant understand a word you say is a big bonus, but make sure you are in contact with a health professional about using any of the drugs I mention before self diagnosing.

I also carry a drug called Doxycycline. It's an antibiotic that is often used as a "prophylactic" antibiotic and I would talk to a medical professional before a journey into the wilds about it. It can cure a lot of stuff, from skin infections to tropical illness and STI's. Its handy stuff to have, and if you have a bad cut beginning to fester having this handy can make a big difference to your chances of getting out alive. Doxy is also a suitable treatment for a tick bite where Lyme disease is a real possibility, with a two-week course "please seek higher medical advice first" upon a tick bite becoming red making a big difference in your recovery.

Doxy, medicated body talc and probiotics are three things I swear by as the best bits of preventative medicine there are. Medicated talc everywhere you sweat is just a must to avoid skin irritation when you could be going weeks without a shower. It's worth carrying a tube of hydrocortisone 1% or better yet 2% cream; it's very good for treating skin infections and rashes particularly at the beginning of a long ride. If I have to choose between putting a water purifying tab in a cup of creek water, or drinking water from a clean flowing stream knowing my body is full of good bacteria. It's a no brainer. If I do get sick, which is very rare, Doxy is the answer. Hand sanitizers, water purifying tabs etc will only strip your body of the natural flora and fauna that keep you healthy in the field.

Powdered sports drinks or oral rehydration salt type sachets are very valuable to have in the case of dehydration, heat stroke and both diarrhea and stomach upsets. Keeping yourself healthy is often as much about prevention as it is cure, and staying correctly hydrated with a gut full of good healthy bacteria is a great start.

There's no such thing as a dumb question.

Ask others, ask us over email or facebook and remember that the only dumb question is one you don't ask! If your serious about getting out there, you can, and you can do it easily and safely in comfort if you get good suitable gear and know how to use it. The best way to know how to use it is to practice with it, and if not sure about how it all works then ask someone who knows and can walk you through it.

But my biggest pro tip... Chap stick. That and sun screen are the difference between smiling on the steppe, and spending a day looking for somewhere that sells chap stick.

Chapter 8.
Saddles, Girths and Tack.

"It is essential to have good tools, but it is also essential that the tools should be used in the right way."
Wallace D. Wattles

When I started this ride, I knew almost nothing about saddle fit, and now as I type I realize that this is the most important chapter we will write. Hell, I adjusted the gullet in a Wintec to fit my fat little mare by pulling it out and beating it with a sledge hammer. I learned a lot since then, and there's three things I wish I knew then that I know now... wait. Stop right there Pete, there's three things I've become to understand are essential: I've learned a great deal on these subjects through my own trial and error, and every day I continue to refine and learn.

To ride a horse on a long ride then it is essential you understand how to correctly fit a saddle, how to balance in the saddle, and overall weight that a horse can comfortably carry. I'll get to the second two in the next chapter, but first we really need to start with saddles, gear and how to fit it. Man... I'm just shaking my head thinking back over all the mistakes we made, and how we've come to be using a mix the most primitive innovations we've learned from ancient horse cultures, and Australian, Kiwi, or US sourced gear.

Saddle fit.

Saddle fit is a topic I will leave for the most part, to the experts. I am not an expert. Further to this I will say that at any point I have questions regarding saddle fit I use Facebook, and consult many of the helpful saddle fit groups who have many expert saddle fitters among them. I will point out a few of the basics, and I could even talk about what size saddle to buy for your height etc, but I won't. I won't because when you buy a saddle from a saddlery, they will know more than me, and if you're taking a long ride with your horse you will no doubt have a saddle fitter chat with you beforehand.

If you're a crazy adventurer like us, you'll be buying a second-hand military saddle from a guy your chatting with via google translate and he will be showing you what he and other nomads have done for thousands of years so you'll be fine; uncomfortable... but fine. I will just cover the basics and talk about what we did, and why were so happy with the result. But trust me on this, if your saddle doesn't fit your horse... you're going to hurt it, and fail at the most important part of horsemanship. Empathy.

To see if your saddle fits your horse, first be sure you are placing it in the proper spot. The tree should sit immediately behind the horse's scapula "shoulder blade bone", not over it. This is a lot easier to find on a thinner horse with a summer coat than a fat little Mongol fluff ball pony. To find the scapula, walk beside your horse with your hand on the shoulder with someone leading him. As the horse moves, the scapula rotates about its axis, and you should be able to feel and see it. Place your saddle "without a saddle blanket" just behind the scapula and girth it sufficiently to hold it securely in place, and keep in mind that its perfectly ok for a bit of the saddle leather to be over the shoulder, so long as it's not impeding movement at all. With the saddle on the horse's back walk around and check it out from all angles.

Most saddle-fit problems involve the withers, "the ridge between the shoulder blades" and with my saddle I like to see a nice clear channel from the withers right through to the back of the saddle keeping weight off the wither and spine. It should fit snugly against the horse and not require padding to fit well as padding out a horse is about as good as wearing 5 pairs of socks in shoes that are too big. A poorly fitted saddle can cause all sorts of issues for the horse, but a few signs to look for are unwilling to take the saddle, soreness, weak hind quarters and general bad temper.

Both Luisa and I groom our horses before and after saddling, and use this time to run our hands over the horse's body looking for tenderness or pain usually given away by twitching, moving away from our touch or a sharp turn of the horse's head towards you with a "mare glare". A horse will often twitch a little, and you can gently put a little pressure on it to see if it's in fact in pain. Horses are very good at telling you if they don't feel great, you just have to listen.

Another very simple trick we feel is essential for a long day in the saddle is the way in which you place the saddle blanket on the horse. Once you place the saddle on the horse over the blanket, before doing up the girth place a couple of fingers under the front of the saddle blanket and sort of shuffle the saddle about as you lift the saddle blanket up into the hollow area of the saddle that runs down its spine.

By doing this when you place weight on the saddle the blanket doesn't draw tight across the horses withers potentially causing discomfort. With the saddle blanket pulled up the withers have room to move and there is no pressure point or "hot spot" so to speak particularly when trotting. This only takes a second or so to do, and is essential if you wish to do all you can towards preventing saddle sores of any kind. Anything and everything we can possibly do to prevent a horse going lame must be done each and every day, and little tricks like this may only make a small difference, but are absolutely worth paying attention to when you rely on your horse every day.

Choosing a saddle for a long ride.

Starting out my first saddle I owned was Wintec synthetic general-purpose saddle I got for free when I bought a mare as a paddock mate for my partners horse.

It was ok I guess; and was my go too saddle for hooning my little standard bred mare up and down the beach, and through the forestry blocks near my home in Dunedin New Zealand. Luisa rode in Western saddles growing up in Bavaria and loved them, but right from the start we both agreed on one thing when it comes to saddles: Australian stock saddles are our clear favorites.

I grew up in Victoria Australia, and school camps or just riding with mates around their farms, stock saddles were the go too. When Luisa did her time as a drover in Australia, those long days with Jill at her heals were spent in... you guessed it; an Aussie stock saddle. So why do we rate them? Why do I ride Aussie stock and what are the alternatives?

Australian stock saddles are designed for a specific purpose, and thats long days in the saddle. Droving, mustering, and covering distance over a variety of challenging terrains often with your swag "Aussie bed roll" on the back and saddle bags on the side. To those that haven't worked with stock before, it may not be exactly what you have pictured. The days are long, and the Aussie stock is as comfortable as it gets for just mooching around. You don't gallop often, moving with the stock often at such a pace they graze happily while you dawdle along with your team chatting about the footy, how badly your keen to sink a cold one, and how dam sick of these bloody flies you all are.

But when you need to move in an Aussie stock saddle, you can move! Polo bolsters mean the horse needs to be just about upside down to get you out of the saddle, and there's a horn to hang off or rope too on most older ones. It's got rings and tie points for breastplates, rear girths and of course your swag and saddle bags. It's just perfectly set up for long days in the saddle, but more importantly, daily life in the saddle; every day... every single day as you move with the stock.

So, is it better than a western saddle? Maybe? But I think "it's different" is a fairer statement to make.

Personally, I have found the way an Aussie stock saddle fits multiple horses of different sizes and shapes to be its greatest asset to me. Mine is an older saddle, and is padded in such a way it moulds beautifully to a wider range of horses than other saddles Iv'e owned

But Luisa didn't have a saddle with her, so we made do with some pretty crafty modifications to a Mongolian saddle that I had planned to use as a pack saddle. To be honest the common saddles you will find in Mongolia are just basic Russian style military saddle, and they can be made to fit pretty much any horse with ease. You'll see them from one side of Asia to the other, and right through into the Caucuses. They fit well, are light, and I can't honestly say they are even remotely comfortable. It's like sitting in one of those metal framed plastic chairs from primary school compared to a lazy boy when you go to one after riding in an Aussie stock saddle. But... they work, and they work well.

Another great advantage of the stock saddle is it can often easily be repacked with wool and shaped to better fit your horse simply and easily yourself.

Although you can count on finding a military type saddle anywhere on the steppe for a very affordable price, you need to be aware that as you travel across the steppe, the size and shape of the horses changes a great deal. As a result, so do the shapes of the saddles, and the way the riders use saddle blankets and pads to best fit their horses. Luisa still uses a military style saddle, but not the one she started with in Mongolia; she now uses a far more suitable Kupkai style saddle we purchased in Bishkek, and much like both my Australian stock saddle and her Mongolian one before, we have heavily modified it.

English, jumping and general-purpose saddles are great in the arena or around the farm, but for both of us, we don't consider them a great choice for a long ride. If this is what you feel most comfortable with, go with it!

I have an "Impact gel" pad for my seat, and both it and the entire stirrup leathers are covered in heavy sheep skin padding. Luisa's has a thick woolen pad too, and again this is covered with sheepskin. Having a warm padded butt on the freezing steppe is great, and it's really not too hot at all riding in summer even with the thick wool; and with the stirrup leathers covered too, I can comfortably ride in shorts without pinching my legs between the stirrup leathers. Both of us would strongly recommend covering your saddle seat if you plan on covering any real sort of distance for many reasons, but in particular it reduces friction, saving your skin from irritation, rashes and time off to recover.

Regardless of what you ride in, make sure its a goof fit for both you and the horse with good points to secure gear too on both the front and the rear.

Blankets, girths, bits, bridles, head collars and hobbles.

Tack is another choice that many people feel needs very careful consideration; personally, I really don't think it needs that much, but there are some basics that really do matter.

Having a saddle that fits is what matters more than anything else, and the blanket you use is one of personal preference. But in short... wool. Anything synthetic is rubbish in my opinion, and for me felted wool is the only option. Wool is natural, and the way it behaves when it's wet and exposed to heat is a result of millions of years of evolution, not lab experimentation with fibers created less than 100 years ago. Cotton is what it is, a fibre that isn't as good as hemp, providing little warmth, it wears out too fast and the properties of all this flash, so called miracle fibers and game changing saddle blankets equate to noting when your using them as your bed each night.

On a long ride... your saddle blanket is going to be a hell of a lot more than a fashion accessory, it's your bed. Wet wool will keep you warmer than a flash bed roll and keep your horse warm, and comfortable during a ride.

The style or make of your saddle blanket is your call. An old woolen military style blanket folded up works just fine, so does a western style one and in my opinion dam near nothing beats a felted woolen rug purchased from a steppe horseman. I've seen these guys literally wrap themselves in these blankets after a day's ride and sleep in a rock crevice during a snow storm.

Sadly, the best one I ever owned was a gorgeous felted Mongolian one I lost when our pack horse bolted, never to be seen again. Wool does get heavy when its wet, there's no denying it, but for the most part it's covered by your saddle and your gear when it's raining, and should you choose to dry it by a fire, it's far more fire resistant than other fabrics. As I covered earlier in regards to our sleeping gear, it's never going to be directly against our sleeping gear and the fact it will keep your warm regardless if it's dry or not, is absolutely lifesaving.

So, with wool against our horse under the saddle, that of course fits as well as is possible; next comes what is arguably the weakest link in all of your gear. The girth. Too tight a girth can make your horse uncomfortable. Too loose makes the saddle unstable, and with the kind of riding we do tighter is often needed, especially in the hills. Girths have caused us the odd issue along the way. and I'm glad they have, because we have really had a chance to find what works.

We use a combination of both modern and traditional methods, that offer both rock solid reliability, and most importantly comfort to the horse. Right from the start I used a mohair girth I paid a good sum of money for. American made, and a mix of wool and cashmere it's nice enough to the touch to be a woman's winter garment. It's almost 20cm wide directly under the horse's brisket and very strong, with perfectly crafted stainless buckles at either end, and was originally secured by a long thick leather strap that ran over the seat of the saddle like a usual western or Aussie stock saddle. I'd been assured at the time of purchase that it was absolutely top-quality leather, and there wouldn't be an issue; but as mentioned, it snapped on me and resulted in yet another broken bone.

To be honest, it was as much my fault as it was the failure of the gear. I often ride with a second girth over the top, and I'd used it to secure gear to the pack horse that day. Lessons were learnt, and I ended up using a bit of an old seat belt to limp through for a while before refining the way I cinch up to its final evolution after our time in Uzbekistan. Our time with Sobir was without a shadow of doubt one of the most important times in my growth as a horseman. Why? Because Sobir could walk into any stable, in any country, on any continent and be the best horseman they had ever seen.

Although a wild man born of the steppe, he was in no way out of touch with modern ideas and materials; it was Sobir who taught us how to girth a horse for the utter war that is Kupkari. The horses and men that play this ancient game are often killed, and there is utterly NO place for gear failure. A buckle can bend or break, a hole in a leather belt can tear out, and there is no room for failure as you gallop at 60kph on a stallion holding a 70kg sheep corpse with your leg among a sea of flying whips, grabbing hands and thundering hooves.

My mohair girth is now secured with a thick seat belt webbing and a synthetic seat belt webbing cinch secured with the same knot you would tie a tie with. There are no weak points, it cannot come undone, it is easily adjusted and I could use it to tow a truck out of a bog. The materials it's made of are easily sourced, if needed it is easily repaired and furthermore its cost is tiny. But its biggest benefit is what Luisa refers to as a "panic knot". The girth can be undone in a split second by ripping on the tail of the cinch releasing the saddle from a horse in an emergency situation.

Sure, a girth like this won't suit your English saddle in an aesthetic or practical sense, but it works perfectly on a Western, Aussie stock, or military saddle, and I use it in conjunction with a second girth that has since become utterly indispensable to me and can be used in a multitude of roles. It is nothing more than a long-braided belt of Mongolian raw hide covered in a Mongolian sheep skin sleeve with a crude, yet seemingly unbreakable steel buckle crafted in some dark corner of the Ulaanbaatar black markets. I've carried this girth with me over more than 10,000km and used it on at least 50 different horses, and would be lost without it.

The steppe horses vary a lot in size and shape, and the terrain over which we have ridden has varied even more still. Although my stock saddle fits a fat little Mongol horse beautifully, without a second girth securing it at the rear, the saddle constantly slips forward on the steeper hills. The large brass loops on the rear of my stock saddle perfectly secure the girth in place when run over the seat, and these two girths are absolutely perfect for work on the steppe.

On the larger Kazakh and Uzbek Karabair horses, this second girth wasn't needed to stop the saddle slipping forward, and was used as a cinch girth directly over the main mohair girth for added security. When we began encountering the steep terrain of the Caucasus mountains, this second girth again changed position to rear of the saddle where its use is now to sit under the horse's tail, and again prevent the saddle from sliding forward. It again sits across the rear of the saddle and is held in place by the brass rings on the saddle and loops in a figure 8 across the rump and under the tail.

I use a breastplate as well, and if available to me I would use one made of mohair. But my access to an item such as this is somewhat limited given our location. I have instead made and evolved mine out of braided para-cord and have padded ever point where friction can occur on the horse with lambskin. It can easily be adjusted using seat belt type webbing and strong plastic clips that can have it on and off the horse quickly. These are more than adequate given it doesn't take a large amount of weight, and I choose to use para-cord as it can easily be used in an emergency situation for another use.

Para-cord is handy stuff, and can be used for anything from building a shelter, raft, snare, or securing gear to the saddle if some other part has decided to let go. A good horseman always has a pocket knife and a bit of hay bale twine in their pocket, but bale twine isn't that common on the steppe, so having a good stash of para-cord up your sleeve is a life saver.

A bridle is something that is completely up to the rider. English, western or any style for that matter work but it's the bit that really counts. I personally rode in a boring old cheap English bridle I'd removed the nose band from until it was stolen in Georgia; I was gutted, as my sentimental attachment was so strong to it and it just felt perfect in my hands.

I've since made a new one out of raw hide in an almost identical manner to the Mongolian nomads. They use a simple raw hide bridle with a nose band that allows the bridle to become a head collar by slipping the bit out of the horse's mouth allowing, it to graze and drink with greater ease. My reins are only secured on the left side too, with a simple hitch tying them on the right meaning I can in seconds turn my reins into a lead rope allowing the horse the greatest possible comfort on short or long breaks.

Raw hide has a draw back though as it can irritate the horse at points where you tie a knot in it, and as a result mine has felted wool stitched in every place it may rub causing the horse discomfort. But rawhide has a ton of benefits, and the lack of buckles on the bridle mean breaking it is pretty much impossible. My reins are also simple and indestructible as reins go, made from braided "purple" para cord... get it. Purple reins... sigh. I am a special kid aren't I.

The bit you choose to use is completely up to you, and what you're used to, but I absolutely swear by and egg butt French snaffle. Snaffles are soft on the horse's mouth, pretty generic so won't be a surprise for a new horse used to the harsh iron bits found throughout most developing nations, but have enough force to let a horse know you're in charge when you want to pull it up. But the large egg butts or rings are the real-life saver.

I've had a panicking horse literally pull smaller bits through their mouths leaving me with no control, and the results normally aren't great for rider or horse alike. Big rings or egg butts are literally a life saver, and if your new to riding or planning to have a go at an adventure like ours, I would strongly recommend a 4 1/2-inch egg butt French snaffle. I've found 4 1/2 inches to be a pretty generic size for steppe horses and the size simply refers to the width of the horse's mouth. If your planning on riding bigger horses, a 5-inch bit could be a handy addition to your kit.

Head collars are something that we got wrong to begin with too. I was struggling to keep the weight down for my 23kg baggage limit and decided to purchase some in the black market on my arrival in Mongolia.

The Mongolian raw hide head collars are as strong a head collar as you will find anywhere on the planet, but they do have one drawback shared by rope head collars as well. They have knots, and wherever you have a knot or any raised part for that matter, you will have a point of pressure on the horse's face. Head collars from your average horse shop in the west are made to be popped on a horse so it can be led out of the paddock and tacked up, but little more than this. But we need to tie up each night and the horses often have their head collars on for as much as 16 hours a day, meaning they need to be both very durable, and also gentle on the horse.

The bridge of the horse's nose and cheeks are the points where the head collar really needs to be as kind as possible, and once again it is wool that we choose to use to protect the horse. Thick felted wool covers these tender parts because as we keep saying again and again, show your horse how much you care for it, show it empathy and look at your gear as though you were the one having to wear it.

We hand stitch our head collars from local materials, and there is another addition beneath the horses chin you won't find too often in the west. We have a metal swivel that prevents the horse from getting too tangled up at night. The rope we use is a bit unique too, and we choose to tie up with about 10 meters of strong, thick, flat rope which is the same as what we use to construct our head collars.

We use this over synthetic or hemp ropes as if the horse becomes tangled in the night, it is far less likely to cause the horse injury and this rope is available in literally every single market and hardware store in Central Asia. The tie up rope is secured by a metal spike with a loop on its end, and a carabiner clip. You can find them pretty much anywhere on the steppe, and the ones in the black market in Ulaanbaatar are particularly good. It's amazing how well they can secure a horse, particularly when the horse is used to being tethered. If there's suitable trees about, we just unclip the spike and loop the rope around the tree and secure with the carabiner.

But there are wild horses all over the steppe, and at times they can disagree with yours at night. We rode stallions for the most part, and a mare in season will always arouse interest in even the best-behaved horse. Hobbles are the answer, and the best hobbles we have ever found again came from the black market in Ulaanbaatar. People often consider hobbling a horse cruel, but it's not; it's anything but cruel and it has some huge advantages for your horse. We hobble three legs: the front two and the rear left, and will often hobble over a short break leaving the horses free to graze as they please.

The horses become far more comfortable with you handling their legs, feet and in general, and very quickly become accustomed to being hobbled. But the real advantage of hobbles comes in the event of a potentially dangerous situation for both the horse and yourself. A horse used to being hobbled and grazing on a lead rope will often become tangled around a bush, in its own rope or around a tree. When this occurs, they just stand there and mutter at you until you come and sort them out. They don't struggle, panic or harm themselves and just quietly cooperate with you as you untangle them.

The saying goes, there's a lot of ways to skin a cat; but the fastest and easiest way to skin any small game is a small cut in its back, you put a couple of fingers of each hand in, and pull outwards. I've done a lot of skinning, and it took me a lot of practice to get things to the point I can have a rabbit skinned, gutted and roasting on a stick over a fire in minutes, and Luisa and I both feel that the above suggestions are solid.

They aren't the only ways that will work, but they are ways that do work very well for us, and I am very confident that should you try them yourself you will have a great start point to begin refining your own style from. A road map will only get you so far, the way you travel that road is up to you; but what we have detailed here is exactly what I would have loved to have been told the day we conceived this crazy journey.

Chapter 9.
Balance and total weight limitations.

"You don't break these animals, you come to an understanding with them."
Phil West.

This chapter is key. It's vital for the sake of your horse and your relationship working over any length of time.

I have asked more questions online, read more studies, scientific papers, talked to vets, endurance racers and trainers about this than any other topic we will discuss in this book. I've even read cavalry manuals from as far back as the Civil war, in an attempt to better understand both how much weight, and how to carry that weight on a horse. Iv'e watched old movies, searched old pictures and discovered in many regard that packing a horse has sort of become a "lost art" in many parts of the world.

Older resources are valuable as the German army of WWII was primarily horse drawn, and in WWI over two thirds of all material sent to the front was hose fodder. For thousands of year horses were our primary means of transport, and in less than a century they have gone from pulling carts, to riding in horse boxes and trucks and all that knowledge became lost almost over night. So much so that now to even talk about the weight we put on horses becomes a touchy subject, especially online.

Sadly, I've caused a lot of upset online unintentionally as a rider's weight is something you simply can't comment on without having the label "fat shaming" thrown around. It's sad really, as all I want to do is understand how people justify the weight a horse can carry and what are the physical effects on the horse over time. I will start by saying right from the start... this is a worm hole and once you go down, things get very complicated very fast.

As a result, Luisa and I have decided to make a very clear disclaimer right from the start, and assure you it is not our intention to insult or upset any larger riders out there.

Our conclusions we draw here are directly related to "Long Riding", not the odd little pleasure ride on the weekend. A horse can take a lot of weight for a short time, but day in day out the damage on a horse can be extreme. Online I hear excuses all the time such as...

"but there's 100kg guys on little cutting horses so I'm fine."

Well, the truth is those guys are balanced athletes on 8-year-old horses bred and built to take that kind of work and your little OTTB aged 4 isn't. But in our opinion "woke culture" has no place in horsemanship. Empathy towards your horse, and treating horses with respect, kindness and understanding supersedes your feelings.

If we are all completely honest with ourselves about what we really weigh in our gear heading out to ride, what our saddle and all our tack weighs... many of us will find we are in all honesty too heavy for our horses to carry comfortably and there are areas we can all improve. Luisa and I both eat only 2 meals a day and work very hard to keep our total fully loaded ready to ride with ALL gear no more than 105kg.

We should never justify putting too much weight on a horse, instead we should strive to offer the horse the best version of ourselves and put as little stress on our partners as is humanly possible.

Again, nothing we say is intended to offend or upset any reader, nor do we make any judgment on the choices you make in your own life and stable, we are simply sharing our thoughts on how best to respect these beautiful creatures that enrich our lives daily.

Balance.

Balance falls into two categories in my opinion; the balance of the rider, and the balance of the weight the horse is carrying on the saddle. The goal of course being to spread the weight as evenly as possible across the surface in connection with the horse and without things flying about all over the place throwing the horse out of step.

Imagine running with a back pack on with the straps at different lengths packed unevenly with a pair of 2lt milk bottles tied to your waist at their handles by string... does that sound fun? Of course not, and it doesn't take a genius to figure out just how quickly that runner is going to start having issues. But let's re pack that bag carefully, place all the softer stuff against the spine side of the pack and put the heavier stuff including the 2 milk bottles up near the top of the pack across the shoulders, tidy up the straps so they are even and do up the packs waist band before running.

Fortunately, I have had instruction on balance from two master horsemen and I will repeat that advice as best I can in their own words here. Let's start by talking about how a rider should be properly balanced, and for me personally there is one man who has taught me more about balance than any other. I call him "Old Guy", and he is a Mongolian nomad whom I have come to respect as one of the greatest hunters and horsemen I've ever known.

On horseback, regardless of speed, and no matter what the terrain is like he simply doesn't move in the saddle. His knees aren't what they once were, and he hates dismounting for anything short of the steepest downhill treks. I asked why he was so incredible to watch ride, and through the rather large language barrier I was told it's because he perfectly distributes the weight between his butt and his thighs, never placing excessive weight in his stirrups. He said the stirrups are not there for you to stand in, but there for keeping your balance and should never take more weight than what you would need to crush an apple.

Old guy has lived his whole life in the saddle, he's killed 8 bears in self defense, and many hundred wolves. He is an enigma of sorts, and spiritually, he is one with horses and the wild.

Back in New Zealand, I was incredibly lucky and privileged to spend some time working alongside a master horseman by the name of Enzo Ferrari, a former rodeo champion from Paraguay. The guy was amazing, and daily I learned things I could never have learned from a conventional teacher. In particular, he taught me how to read body language and how to touch a horse.

He knew all of this from his rodeo days, and would know which horse to pick and just how to piss it off in such a way he got that winning ride, the belt buckle, the dollars and then of course; the ladies. Enzo again mentioned stirrups, telling me to take of my stirrup leathers and measure the hole exactly, and I do mean EXACTLY. Hang them on a pair of nails from the buckle and make sure there is not even a millimeters difference between the two, as over a few days riding this is enough to throw both you and the horse out. I was skeptical at first, but he is absolutely right.

His thoughts on stirrup length in relation to balance were also very insightful and he told me to slip my feet out and push my feet heel first towards the ground. If the stirrups aren't hitting your ankle bones, then adjust them until they are. I've ridden this way ever since and on many occasions been told primarily by English riders to shorten up. Nope, this is how the Mongol's ride, the Paraguayan rodeo champion told me to ride, and it's how I personally feel most comfortable in the saddle. Enzo also never gave me a hard time about the way I sat in the saddle, as I tend to slouch a bit, again something an English riding instructor would take displeasure in viewing, but as long as you carry the weight evenly, both you and your horse will be riding safely and fluidly.

When it comes to the balance of the gear you carry, the simple rule is this, keep it as even as possible front to back and side to side. Old westerns are a good resource, and weight is carried usually in a pair of saddle bags either side of the saddle, with a sleeping roll across the rear of the saddle. Across the front of the saddle it's coats or blankets draped over the withers in a roll, with heavier items such as rifles and in particular canteens hung again at the front of the saddle on either side. The biggest benefit of a set up like this... You can gallop, and it is this set up Luisa and I have come to adopt to make riding as enjoyable as possible for us and the horses.

Lu carries our sleeping bags, the laptop and little bit of clothing in her water proof roll bag, and I carry our three-man tent in mine. These are our softest items, and they do at times touch the horse's spine area no matter how hard we try to prevent it. But as these are our softest items they don't rub, bruise or hurt the horse.

We have both sewn drink bottle holders onto the front of our saddles too, and can easily carry two 1.5lt bottles on the front of our saddles. I also like to keep a metal flask for vodka, chacha, or whiskey there too, "simply to offset the weight of my ropes opposite of course". We've also sewn on some webbing type straps with clips so we can easily keep out wet weather gear and coats across the front of the saddle, and grab them when we need them. On either side of our saddles at the rear is where we keep our food, pots, stove etc and we are both now very happy with both the balance, and evolution of our saddles and the way in which we carry weight on them.

How much weight can a horse take?

I was always told 20% of the horse's body weight is the magic number it can take on its back, and this number comes from the old cavalry manuals based on horses in great condition soldiers relied on for their very lives covering a good 60km a day. New veterinary study on pleasure horses suggest 15% is now days a far more realistic number given the way in which pleasure horses are used so differently to the cavalry horses of the past.

But a study of 374 competitive trail riding horses compared horse/rider weight relationships during the running of the Tevis Cup, which concluded that these horses can easily carry over 30% of their body weight for 100 miles and not only compete and pass vet checks but compete at the highest standards. Breed of horse, age, it's condition score, muscle condition, bone structure and fitness are all factors that need to be considered and fortunately there have been a ton of studies done on all these; unfortunately, the conclusions vary so widely I just ended up more confused than I could even imagine. But after a great deal of research I think I am able to draw a pretty reasonable method of determining how much weight your horse is able to comfortably manage.

Breed of horse is a big factor, with smaller stout breeds like the Fiords and Mongol horses renowned for their power and strength despite small size. Bigger horses, such as draft horses are bred to pull, not so much carry weight and despite being far larger can carry proportionally less. Age is a big factor too, with a horse's bone structure not reaching full maturity until 7, so it would be my strongest recommendation to avoid considering a horse under the age of 7 for a long ride.

Condition score is the next big thing to consider as if a horse is overweight, that's weight it is carrying before you even factor in you and your gear! The scary thing is a huge percentage of pleasure horses are overweight, and many of them have owners that don't even realize it. Then of course there is fitness, and a horse that mooches about in a paddock all day is far from ready to tackle a long ride regardless of the weight you put on its back. So, after all of this meandering about what do you need to know to make a sound decision on purchasing a horse for a long ride, particularly if you're in a far-off corner of the world with little ability to communicate.

My formula is as follows. First, make sure the horse is between preferably 6-10 years of age then work out its weight. There are a few ways to do this, with scales of course being the most accurate but unlikely to be found out on the steppe.

You can use a commercially available weight tape, or measure the horse with a regular tape measure by calculating; heart girth X heart girth X length, divided by 330, + 50 = weight in pounds. I find, having tested both, and getting actual weights on scales while learning as much as I could prior to heading off on our journey, that the weight tape is usually a bit less accurate than the tape calculation, and both are usually a bit light compared to actual weight. Knowing your horse's weight is important not just for figuring out how much it can carry, but it's also important for giving the correct dosage of medications, another thing we've learned to do ourselves on the steppe.

There are a couple of other methods that can be found online, but this is the one I use.

Next, I consider condition score, and for me personally I would like to have a horse as close to a score of 5 or just above it as possible, with below 1-3 considered underweight, 4-6 healthy and 7-9 over weight. You can very easily score a horse yourself and get a pretty accurate idea even with little to no experience using the guide I've included bellow. I feel condition score is an important thing to consider as a horse that's underweight and badly conditioned will struggle to carry weight, just as much so as an overweight horse will be burdened by the extra weight it carries before you even add you and your tack, so it's important to consider factoring this in too.

A study done on the horses that competed in the Tevis Cup revealed a condition score of 4-5 is good, 5.5 is perfect and horses that scored higher were far more prone to lameness throughout the race. These are things well worth considering in the selection of a horse for a long ride. Further to this regularly monitoring your horses condition is important for both its wellbeing and saddle fit. Many horses weight will change throughout the year, and wild horses in particular look amazing going into winter, and not so great as the first of the spring grass starts to come in.

You can easily find numerous images of condition score guides online, and I keep one of these saved in my photos on my phone. On the following page is a very good guide you can use to work out a horses condition score, with a quick glance and a prod.

Horse condition score guide.

1 Poor.
Extremely emaciated; no fatty tissue; vertebrae, ribs, tail head, and bones of withers, shoulder, and neck are visible.

2 Very thin.
Emaciated; slight tissue cover over bones; vertebrae, ribs, tail head, and bones of withers, shoulder, and neck are visible.

3 Thin.
Slight fat cover over body; individual vertebrae and ribs no longer visibly discernible; withers, shoulders, and neck do not appear overly thin.

4 Moderately thin.
Ridge of spine and outline of ribs are visible; tail head may or may not be visible depending on the breed; withers, shoulders, and neck do not appear overly thin.

5 Moderate.
Spine and ribs cannot be seen however ribs can be felt; tail head is spongy; withers, shoulders, and neck are rounded and smooth.

6 Moderately fleshy.
Slight crease down spine; ribs and tail head feel spongy; fat deposits along withers and neck and behind shoulders.

7 Fleshy.
Crease down spine; ribs have fat filling between them; tail head spongy; fat deposits along withers and neck and behind shoulders.

8 Fat.
Apparent crease down spine; ribs difficult to feel; soft fat surrounding tail head; fat deposits along withers, behind shoulders, and on inner thighs; neck is large.

9 Extremely fat.
Obvious crease down spine; patchy fat on ribs; bulging fat on tail head, withers, behind shoulders, and on neck; fat flanks and thighs.

Measurement Test

As would be expected, good body condition and bone structure were found to be paramount in Tevis Cup winners, and the studies conducted on those horses have given us a method to determine very accurately the weigh carrying ability of a horse. Bone structure was evaluated using the front leg cannon bones as representative of general structure, with a measurement in inches taken halfway between the fetlock and the knee.

The equation is simple, and below I've used Bill the Bastard and myself as an example. Sadly like many of the equations Ive mentioned they are in imperial measurements not metric, but converting them is very easy, and given we measure horses in hands anyway... it isn't to much of a hassle.

1. Add up the total weight of the horse, rider and tack.

Our example: Bill the Bastard 1010pounds, Pete + tack 210pounds = 1220 pounds.

2. Measure the circumference of the cannon bone midway between the knee and fetlock.
Bill the Bastard, 8 inches.

3. Divide this total weight by the circumference.
Our example: 1220 ÷ 8 = 152.5

3. Divide the result by two.

152.5 ÷ 2 = 76.25

Values near 75 are great, below 75, even better. Values from 75/80 are acceptable. Values over 80 indicate weaker legs and care must be taken particularly in challenging terrain. Values over 85 suggest you need a horse stronger horse, or to ditch some weight.

In this example you can see the weight I asked Bill the Bastard to carry was well within what he was capable of.

A horse can increase its bone strength over time too, particularly if you do a lot of longer, but slow, distance rides allowing the horses overall strength and fitness to build up over time. I mention this because it's something to consider when purchasing horses on the steppe where I would absolutely swear by attempting to purchase a farmer's stock horse. These horses are ridden all day, slowly with the stock meaning often they have built up an excellent structure for long riding. In addition to this, they have usually got the best temperament of the horses you will encounter on the steppe.

Be honest with yourself about your weight. Be honest with yourself about the weight of your gear and be honest about the condition of your horse. Some riders are just too heavy; that's something you can change to a degree, and there is always the option of riding a bigger horse slower over shorter distances each day and taking a pack horse. If your horse and you are overweight or out-of-shape, you both must be conditioned, slowly and appropriately, prior to carrying a heavy load on a long ride.

Such horses should not be expected to carry more than 20 to 25% percent of their body weight, and I would strongly recommend you have your veterinarian perform a thorough "soundness" exam if that is humanly possible. Be honest about the weight the horse will be carrying and ask them to be especially thorough in the evaluation of the horse's back structure and the suspensory ligaments in the legs.

Although pretty much every horse we encountered on the steppe was barefoot, shoes can be an advantage when you start to spend a bit more time on roads or in steeper terrain. I felt mentioning this here is relevant, though not essential, and it really is a case of a horse with bad feet, can't carry weight. A hoof should be balanced, as odd angles or heights increase the stress on its feet, legs and back. Much like doing a medical course prior to departing, see if a friendly farrier will let you spend a day or two with them, shout them lunch and a few beers and get them to walk and talk you through the ins and outs of what to look for in sound feet.

Chapter 10.
Survival, field craft and navigation.

"I went to the woods because I wished to live deliberately, to front only the essential facts of life, and see if I could not learn what it had to teach, and not, when I came to die, discover that I had not lived. I did not wish to live what was not life, living is so dear; nor did I wish to practice resignation, unless it was quite necessary. I wanted to live deep and suck out all the marrow of life, to live so sturdily and Spartan-like as to put to rout all that was not life, to cut a broad swath and shave close, to drive life into a corner, and reduce it to its lowest terms..."
Henry David Thoreau.

Luisa lays still, the toned curves of her back pressing against me as she sleeps. Jills awake, standing up in the tent in the dim orange light filtering through the canvas as the sun rises over the steppe. She takes a step forward, puts her paws together and drops her shoulders in a long, slow, smooth stretch as she opens her mouth and arcs her tongue in a long yawn that climaxes in a little yelp. She shakes her ears, wags her tail and looks towards the opening of our little house on the prairie.

The day had begun apparently, and its first order of business was letting Jill to relive herself; or as Luisa would say "a toilet". Lu murmurs and groans in displeasure as I slip out from under our sleeping bag, allowing a cold wave of air to stroke her bare skin as I throw on my puffer jacket, unzip the tent door for Jill who rudely bursts through the half open zip and disappears into the tall autumn grass. I follow Jill into the pre dawn light, already illuminating the endless rolling steppe in soft shades of amber and straw yellow before the suns first unobscured rays light the sky a brilliant blue.

I join Jill in relieving myself, wiggling my bare toes in the frozen ground as my urine cuts colorful patches into the white frost of the frozen soil outside our tent beside the remnants of last nights campfire. It was surrounded by empty beer bottles and a pot containing the frozen left overs jill was unable to lick free.

Billy calls to me, softly at first; then with a little urgency in his tone. Jac grazes quietly nearby having clearly pulled his rope tether free at some point last night, but billy wants to have a word. My feet burn almost to the point they feel warm on the frozen ground as I walk to my muttering stallion who excitedly drops his head and allows me the freedom to unclip his rope for a morning of free grazing. He whinnies softly; rubbing his muzzle against my shoulder as he pushes me about in a cheeky but grateful manner.

I wander towards Jac, knowing Billy won't go far off his rope with his hobbles still on. Jac, as always, just turns his back and is a total dick to me. I swear this gelding thinks he's a mare, so I grab his trailing rope pulled during the night, and let him know I'm in charge. As always, he bows in submission and shows me the same love the Bastard just did, and is grateful as I unclip him for a morning of free grazing.

My feet are cold. Really cold, but we need water and a brew on so I grab the jet boil and wander toward the creek only meters behind the tent. I probably should do the dishes; but yeah... nah. The sun isn't even up yet, so I wander back and fire up the jet boil, sitting in the doorway of the tent wondering where Jill is as I hadn't seen her in a bit. I slip on pants, brush the frost, dirt frozen to my feet and filth off with the same socks I've worn for a week, recycle them; and chuck on my boots ready for the day ahead.

The waters boiling already, but I've got no idea where Jill is; and decide to drop a large piece of chocolate into our mug, add a coffee sachet and give it to my sleeping beauty before I begin my search for the world's worst cattle dog. The burble of the creek behind me, the clicking of the spoon stirring the brew and the mummer of the horses were the only things breaking that perfect silence of the steppe. I greeted Lu as gently as always, slowly drawing back the sleeping bag she'd covered her head with saying

"Hey ball-bag, brew for brekkie."

But I wasn't greeted by the back, or for that matter front of her pretty face; just a little cattle dog that had snuck back and stolen whatever warmth she could.

Morning routine followed, and I went and caught a few grayling for breakfast. Lu made a fire, and Jill did dog stuff as we tidied up and the boys grazed. Lu really enjoyed the fish, and did a great job cooking them in a bit of foil with some of the ridiculous amount of unnecessary herbs I felt she carried. But man; I felt good. The wild man of the steppe, I'd caught breakfast and now I got that beautiful little crooked smile and "thank you" from the woman I loved.

It was better than paying rent like a normal person, waiting for the weekend, a dinner date, a movie or whatever it is normal people do. It was just life on the steppe, and knowing we had the day with the boys; Billy and Jac, and of course our little special kid Jill... yeah. The so-called real world full of people saying I was crazy just seemed so far away; I felt like the only sane man in a world gone mad. I mean just look at America, Trump and Biden... thats like being told to choose between a Nickleback and Creed CD for a four year road trip

As the tent fell, and Lu began to fold and pack gear, Billy made his way towards us, looking as always for his brush. I removed his hobbles and put on his bridle, brushing him for a while as Jac did all he could to distance himself from us. Jac was a dick, but lovable and perfect in his own way. Amazing to ride, and better than Billy in every way imaginable under saddle, but he was a moody fella; and Billy was my bro.

As Lu tidied the final bits of camp; I threw myself onto my stallion, and cantered off to pick up Jac who was interested in anything but the day to come. Jac of course didn't fight me or Billy; he just gave me his head; accepted his bridal and let me take his hobbles off without fuss for a change.

I stroked his sides, his face, and stood between our boys just soaking it all in as the sun rose, the day began, and life... life as I know it took one of those paw-together stretches with the arced tongue yawn ending in a yelp. I threw myself on top the Bastard again, and bent towards camp and thought forward thoughts, dropped my reins, and willed Jac to follow.

I cantered in; bare back atop a stallion with our stock horse in tow having caught the woman I love breakfast to be greeted by a dog that loves me only because I love her mother more than I love myself. This was living.

We walked a little further that morning than we usually did after tacking up the boys as the stream, flanked by low willows in there Autumn hues were just so dam pretty. We laughed a lot, Luisa making fun of my antics the night before playing around with Billy after a few to many "Lady Penelope's".

The lady Penelope is a name we invented for our "steppe cocktail" of choice. It's basically 3 parts vodka, 2 parts apple juice served in a bowl, and it's named after a character we invented during one of our more ridiculous days joking about. We were freezing that day, so I would always distract Lu with jokes and fun games we could play, and that day I decided that the next people we met I would introduce us as "Her royal highness lady Penelope Myar", and myself as her faithful man servant and protector "Allan Quartermaine", and that we were on a quest of some sorts looking for rare and exotic birds to watch.

We spent most of the day talking in posh English accents, making jokes about life in our castle and the fun parlor games our servants must be having in our absence and yep... you guessed it; this was all over a flask of apple Vodka.

When Luisa starts laughing, it can deteriorate very quickly into her unable to stand, doubled over on the ground in tears in a state of utter hysterics, and this morning was another of those days. She was breathless and in tears at points, trying to explain how I fell off Billy trying to ride him down to the creek with his hobbles still on, and how I lay under him laughing as he just looked at me like...

"Really Pete?"

Billy loved me as much as any horse has ever loved anyone, and although a stallion, he was just the softest, kindest, big old teddy bear, and really loved our cuddles each night.
I'd ride him bareback most nights to water, and often just lay on his back hugging him, talking to him about stuff. Stroking his sides and enjoying his soft murmurings and horse smell under those endless stars, free from the light pollution of normal life. I'd talk to him a lot about how we had become a herd, and how cool it was that our routine had become so familiar for us all it just had us all feeling at peace.
I was truly amazed at how fast he and Jac had accepted this life, and learned to love it. No two days on the steppe were ever really the same, but this feeling of belonging to something; a tribe, a family... I don't know the word for it. But that feeling was magic and it never once left us during our wanderings. And todays wanderings would be the same as every other for our family; a search for good pasture, clean water, and shops selling our staple diet of canned horse meat, canned fish, rice, fresh root vegetables, meat, beer, vodka and kebabs.

Finding food.

We had no idea what to expect as far as finding food went, and we got it wrong right from the get go. We had just way too much food of every kind including a 5kg bag of rice we carried with us as though we were planning to cross Kazakhstan without a single re supply. We were way off, and one thing of note about the steppe I failed to realize, is the distance between towns has been determined by horses.

Chengis Khan and his hoards could only go so far between stops, same as the silk road traders and every horse back traveler that has been wandering the steppe for as long as horses and man have been in cooperation. As a result, small towns dot the steppe, and the distance between them is never so great you feel completely without hope of resupply. Prices are dirt cheap, and Luisa and I never had any real trouble finding food or resupplying, allowing us to travel lighter and faster, with our "emergency rice" going from 5kg to just 500grams.

There were of course times when we carried a little more, and pushed further into the steppe for real solitude and a sense that we were the only people on earth; but even still there was always a road or trail used by Shepard's or nomads we would cross, and waterways would almost always have some sort of farming activity along it at one point or another. It was strange really, you'd be literally in the middle of nowhere and a little white Lada would come bouncing along over the empty steppe, and a bunch of excited and friendly Kazakh's would soon be offering us vodka, sandwiches and a place to stay for the night.

The food you'll be offered when welcomed into homes varies, and your comfort zones may be tested a little. Be aware that horse is on the menu here, and it's become one of my favorite meats to eat. It's got a delicate flora taste and great texture, and is served in everything from dumplings to steaks, and so long as you can emotionally detach from the fact your best mate is quietly grazing nearby as you stir a can of horse meat into your stew, you'll be just fine.

Vegetarians are going to have a hard time, and if milk and cheese aren't your thing you're also going to suffer. Fresh milk from horses, goats, sheep, camels, cows and yaks is available everywhere you go and the local cheeses, yoghurts, and even beers made with these taste like nothing else in the world. The mare's milk beer, is easily distilled into vodka too, and mare vodka isn't that strong, usually around 14% alcohol and it's certainly got a taste all of its own. The beer is much the same, and be warned... hit it hard and you'll know about it from the brutal liquid sit down that follows.

The gut bacteria of those that live on the steppe have evolved over time, and although I never quite adapted to large doses of steppe home brew, the probiotics I take daily helped no end. I recommend you open up to the steppe, and try all the weird things offered too you no matter how much the thought of eating a wolf dumpling, a boiled marmot "the last animal on earth known to regularly carry the black plague", or a glass of fresh camel milk makes you feel uncomfortable.

Honestly; we lived very well on the steppe, and the beautiful vegan chef that has accompanied me for the last year has not only opened up to the idea of eating meat, but rubbish like 2-minute noodles and canned meats. She became adaptable; and for both of us, hunger often dictated what we considered tasty.

Basic survival, keeping safe and picking your camp site.

Surviving in the wild is easy if you follow a few simple rules, the biggest of these is; DONT PANIC! Keeping warm and dry isn't hard with the right gear and a bit of knowledge. Simple things like picking a campsite out of the wind, away from a river that may rise in a coming storm, or under a dead tree with rotten branches that may fall can mean the difference between life and death. Knowing how to make a fire quickly, and how to dry your gear without burning it or allowing sparks to fall onto your tent are the simplest of things it's very easily to assume people know how to do; but many have never learned.

There are literally full grown adults living in cities that have never learned to cook, make a fire or have even the first idea of how to pitch a camp outside. That's not to say they cannot take on an adventure like us; but we would strongly recommend if field craft and basic life in the outdoors is completely new to you, then work up to it with a few camping trips preferably with someone who can mentor you a bit. There are camping and hiking groups the world over and rather than write a ton here, we'd suggest you join such a group and up skill on the basics. Better yet, look up Ray Mears and book yourself into one of his courses if the opportunity exists.

Keeping safe is often as simple as washing your hands properly; COVID-19 has taught us the importance of this, as well as changing the world we live in. Good hygiene in the field leads to comfort in the field. Diarrhea or vomiting is awful at the best of times, and both can very rapidly lead to dehydration and death. Believe it or not diarrhea kills as many as 1,600,000 people a year, primarily in the developing world, and well: that's more than COVID-19! Having the good sense to avoid its causes, the ability to prepare your body physically to fight it before it takes hold, and a bit of medication to treat it is just smart and simple insurance.

Wash your food pots and utensils properly, keep your body clean and use medicated talc to avoid rashes and skin infections. Change your socks as often as you can, wear natural fibers and stay hydrated! People think keeping safe is about dealing with worst case scenarios, but for me, it's not; it's about being smart right from the start and simply avoiding getting into a dangerous spot to start with.

But what about horse thieves? Prairie pirates and corrupt government officials and cops? Look... there's nothing you can do about that. In the developing world there will be risks; be it corruption, a bear, a wolf or a guy with an AK47 that wants your horse more than you, that risk is there and always will be. But then there's COVID19, crime, drunk drivers and a million things that can wipe you out. Out here on the steppe, there is risk; but play your card right and show a little common sense and it can for the most part be avoided without a great deal of hassle.

Avoid dealing with drunk guys in particular, and defiantly avoid drinking with strangers in strange places. Luisa has her rules for solo female travel and they have serve her well, I have my rules too and between us we have both managed to avoid issues for the most part by being careful in where we camp, who we trust, and how we behave in a country with cultures and customs that differ from ours greatly.

We personally like a bit of space in the evenings, and we like to camp usually at least a few kilometers from towns for a couple of reasons. Shepherds are heading home as we head out, and we can usually find somewhere nice and out of sight to pitch a camp before the sun sets. Shepherds will almost always want to invite you in for the night and as good as this is, often avoiding social obligation is preferable for us. Being a few kilometers out of town means you will probably meet the Shepard's the following morning as you tidy up, and they always want to chat which is great, but can take up a lot of time when your often keen to get moving.

We like to camp wherever possible in a fold of land where we are hidden from roads, tracks and main stock routes in order to reduce the chances of being harassed by a drunken opportunist, or a bunch of rascals that to be completely honest; can be found anywhere in the world. Be polite, be professional, and have a reserved degree of trust for everyone on the steppe just as you would at home.

But... never forget the most dangerous thing is most likely your horse, and complacency can kill. Horses bite, and they can literally kill you with a single kick. I will always talk softly to a horse no matter what I am doing, especially if working behind it so it knows where I am and is not startled by my sudden appearance as this is about the best way to get kicked. I also always like to keep a hand on the horse no matter what I am doing around it as if it turns to bite you; you can feel it through its whole body and can get out of the way and give it a stern reprimand.

If I am going to walk behind a horse, which I often do, then I walk as close to the horse as possible with my hand on its rump at all times so it both knows I am there, and if it does decide to kick I am so close to it that it's almost impossible for the horse to kick you with any real power. About 1m or so back from a horse is about the perfect range for a fatal kick, so either out of range completely, or so close it can't hurt you badly are the only safe places to be. Never forget no matter how well you think you know your horse; they can still do stupid things for seemingly no reason at all; Beersheba being the perfect example of this.

Feeding the horses.

Horses need a good 7-8kgs of food a day if you listen to some folk, others say they need a balanced diet of supplementary feeds, oats, minerals, hoof supplements etc etc. Well, on the steppe your horses eat what's there, and for the most part it's a mix of herbs and grasses; the horses work hard and we let them graze as much and whenever they want so long as they aren't dictating the pace of the ride.

If we get onto good grass by a river, we stop and have a beer, a snack and give the boys a good feed; and every lunch time we aim to stop on the best grazing we can find for at least an hour of free feeding. At night, it's the same thing; we try and find the nicest possible feed we can and move the horse's tether's last thing before we go to sleep so they have as much of a chance to feed as possible before moving or allowing to free graze again the next morning. When you're with your horse 24 hours a day, you generally get a good feel for when they are happy, hungry and hangry!

But there are times when food is lean, and taking supplementary feed is a burden we aren't really very well equipped to take on. You can't just carry a bale of hay with you when you're only on a horse each; but what we did carry a big empty hay bag, and as we liked to camp "near" but not in villages, we would always aim to try and buy "although its usually gifted to you" a bag of hay towards the end of the day; then just ride off into the steppe with it on our lap.

Grains and barley can be purchased along the way, and the best way to get it, is to carry pictures on your phone of exactly what you're after. Point at the hay bag, point at your mouth and make chewing noises and rub your belly; then point at a picture on your phone of a stack of hay, grain etc and then flip to a picture of a horse. Almost, if not every house on the steppe, has a stash of hay out the back and they will happily help you out or point you to someone who will.

This same technique works for finding the horses water, and most people will happily come out with a bucket for your horses. We both have special photo albums on our phones containing a ton of pictures we can point too; everything from ropes, hay, lucerne, grains, carrots, buckets, horseshoes and basic sentences written in the local dialect screenshot for when there's no internet for google translate to work.

Navigation.

Navigation on horseback is no different to another way of getting around when you break it down, but with the way technology has evolved there's a few tricks worth mentioning so you are not reliant on only one form of navigation. Why? Because the entire assault into Baghdad was held up by a lack of AA batteries, and young officers that had no idea how to use maps and compasses. There is always a way, and sometimes the shortest isn't easiest, safest or practical; and knowing how to choose a route using all the information available to you is critical for both you and your horse's safety.

I grew up in the 80's in Australia, and from age 5 it was my job to navigate for my Mother. My Mother is quite unique. She is incredibly intelligent, but much like both Luisa and I she is different, and part of her brain is "missing" so to speak, with other areas so beyond others it beggars belief. My Mother was a mathematician; and an utterly brilliant one at that. But she had literally no ability to find her way around town, let alone find where she had parked the car when we went shopping.

I vividly remember sitting in the front seat, barely able to see over the dash with the phonebook sized "Melways" map book on my lap, directing her around suburban Melbourne as a kid. Things have changed a lot since then, and as I grew, I learned to navigate through my own adolescence and teen years driving both with the "Melways" and a technique I swear by to this day. If your riding in a group, don't let ego get in the way. If someone has a better idea what they are doing, find a strength that compliments this and work together.

I couldn't drive, mum couldn't read a map, but between us we got there efficiently and without issues through knowing our role and supporting each other.

The old saying "real men don't need directions" is utter rubbish. Petrol cost money; if I was lost I'd stop at the first shop I saw and asked directions. Asking directions is a piece of advice I will give in regards to navigation for several reasons.

The people you will meet along the way know things you can't find out any other way. They know when a ford's been washed out, when there has been a slip that's blocked a pass and where the dogs likely to cause you grief are. They know where good feed for the horses can be found, and where there are risks of wild animal attack, good camp sites and everything in between. But more often than not they will just help you find the easiest and most direct way to where you're going, and get a real kick out of being part of your journey.

Shepherds are very helpful usually, and on several occasions we even had people offer to ride with us to show us the way or point out tracks to take so don't be afraid to stop and ask for help, who knows who you will end up meeting and what advantages it may hold.

Old school maps are another resource that can be used. I learned to read a map at cub scouts, learning to use a compass and doing orienteering courses on both cubs, and then scout camps as a kid. I did this at school too, and even spent a week navigating around a semi desert area of South Australia during a school camp where we were basically set free to wander and find our own way around.

I was comfortable reading contour maps from a young age, then during my time in the military became very comfortable learning how to do everything from plan routes, to calling in fast air and artillery strikes on enemy positions. My training during 2005 was focused only on map and compass work, with only directing staff carrying GPS units; the idea being we learned how to do it "old school" so we could fight regardless of what extra kit we had, or had lost or was unserviceable. So long as a rifleman carried his map, compass and rifle; we could complete our mission.

These days maps are getting harder to get, and learning to use them to the degree I have been trained too is becoming very rare, and although it is a skill I am very grateful to have, it's becoming less and less of a necessity.

GPS units are now cheap, affordable and bloody good. I used a pair of Garmin Rhino's hunting in the New Zealand bush as well as throughout the South Pacific for years. Units like these are amazing as they not only tell you where you are, but can have maps of almost anywhere in the world loaded onto them and double as two-way radios. You can chat with mate's who are nearby or scouting hunting areas, and update your position on their map instantly keeping you both aware of each other's exact location.

If you were to fall, you could easily get help directly to your location by simply messaging your companion. However... batteries last a couple of days at best, and although you can get battery packs that take AA batteries for these they can only transmit on their low 2 watt setting instead of the 5 watt, shortening the range of the radio as well as needing a lot more weight in the pack. A good GPS is an advantage, but believe it or not... I decided not to carry one due to weight and confidence in of all things... my cell phone.

I have an iPhone running a pre-paid sim card that I switch as I enter each new country, and personally I find it makes a GPS obsolete, that and you would not believe how great cell phone reception is on the steppe! You can download any number of map programs, it's easily recharged via a small solar panel or by a power bank, it's also a camera, video camera, music device, tv, game console and means of communication. Maps work offline and you can get both topographical maps as well as... GOOGLE EARTH!

I can actually see what's up ahead and use it for finding clearings to graze the horses, unmarked bridges, unmarked water holes and even farmers water troughs on the steppe! Best part is YOU CAN LITERALLY SEE THE PATHS TAKEN BY STOCK! It has been the best resource imaginable for our journey and I use "Google maps" for almost all of our navigation. Remember most maps can be downloaded and used offline, and your GPS works offline so rather than spending more money on another bit of kit, I strongly recommend a smart phone as your go too.

We both carry them, and we both carry the means to charge them so we do have a backup, both are waterproof, and can give real time weather information, have compasses and even a barometer!
But here comes the big thing either have, or don't have... and it's what I call "bush sense". My mum has none, and after my years in the bush I can keep myself on track with little use of digital aids. If I'm in the bush, I follow rivers and ridges, the natural game trails and the many paths nature carves for us and the other forest dwellers to follow. I know how fast I'm covering ground on foot, and on horseback just by feel, and all of this comes with time spent in the bush.

Horse ride tracking apps are really handy here too, and can give you a good idea of the speed at which you cover terrain in "real world" situations. When you build up a bit of knowledge you can plan how far you aim to realistically travel in a day after checking out what google earth, the locals and the topographical map says lies ahead in the coming days. Second guess yourself, ask questions and most of all learn to trust your compass and maps over anything your head may tell you in bad weather. Take small steps at first, learn to stay safe, and then push further and further.

It would be wrong at this point for me not to mention locator beacons, "EPLB's" and satellite phones that can be both bought and hired. I don't have either. I don't have either because I don't want to carry the extra weight and the places I may need one, there's no one who will be coming to save me anyway. In the US, Australia or NZ where there are great rescue services it's something well worth considering; again, I choose not to carry them.

My reasoning is this; people take risks with their lives when they think "worst case I'll just call for help". Well I can't call for help, and as a result I plan every aspect of my navigation, my evacuation plan and my medical plan to factor this in. I rely on myself, my partner and my gear to get me out alive, not a phone call or panic button that may or may not work. Like I said right at the start, AA batteries held up tanks invading a country and how you plan is up to you; but I use a smart phone knowing I can get myself out of whatever I walk into on bush sense and a paper map alone if need arises.

Missing.

But what if you do go missing... who will know when to worry and who will even begin to know where to start looking. Well; you guessed it, again it's smart phone to the rescue. I have what's called the "lost communications procedure" in place with both Luisa's family and friends and mine.

Both of our phone's locations can be accessed through the "find my phone" app, and we have given the sign in details to multiple people who can first contact each other in the event of concern, and then locate the last know location of both phones. We have both got designated people who will know when we are overdue, and when and if too worry. If I'm going to be out of reception for a few days, I will let them know and give a date and time when they should begin to worry. First thing to do would be to try Luisa if I can't be contacted, then her point of contact.

If contact is lost, we are overdue, and people are worried they will know where we were, where we were headed, and the last known or current location of both of our phones. It's a simple and effective method that can give your family peace of mind, as well as potentially save your life in an emergency.

Hunting, fishing and foraging.

I'm watching "Meat Eater" as I type this morning, and Steve Rinella is a guy that just "gets it" in regards to man's place in the natural world. He, is an inspirational modern-day outdoorsman and his show on Netflix is about more than hunting and fishing; it's about a connection with the environment and your food, and showing harvested game the utmost respect by preparing them to the best of your ability. It's about time in the wilderness with like-minded people, it's about seeing things others don't and working to overcome adversary. But when you succeed; particularly in the most challenging of environments, that meat tastes so dam good and you and your quarry are connected forever.

Even if you've never hunted before, if a chance to be part of a hunt presents itself in your travels, I would highly recommend joining to experience the highs and lows of harvesting wild game. If you're interested in learning to hunt, a mentor from a deer stalking or hunting group online, or joining some sort of hunting club is a great place to start. Many run club hunts, and these are a great way to learn the basics and in particular learn the art of processing an animal after you have harvested it.

Be aware though, that life on the steppe will always involve the killing of sheep or goats during your visit to remote communities, excited to show random horseback travelers the highest degree of hospitality. Again, I would strongly recommend watching how in particular the Mongols kill an animal for human consumption. They make a small incision in the stomach, reach inside and tear the arteries from the heart in a very quick, relatively painless procedure in which no part of the animal is wasted.

No blood is spilt, and every single part of the animal is used with even the last remaining contents of the stomach placed on the garden. The meals that follow are amazing, with vodka, beer and a warmth of hospitality that makes saddling up a horse to move on the following day a sad affair with a heavy heart and often a sore head. It's a good way to learn how to skin and butcher your own animals too, and even if you're not prepared for a hunt you can easily purchase a young goat or sheep for your own needs.

So, do I hunt much? Not overly; there's a ton of laws that stand between me and legally harvesting wild game on my travels and I abide by these. Indigenous hunters have a different set of rules that apply to them, and on many occasions I have had the honor of hunting alongside them on their hunts, taking deer, wolves and other wild game. On these occasions Iv'e usually used their rifles "usually an ancient AK47 or WWII era bolt action rifle with crude iron sights", or my bow. But I don't carry a rifle for a couple of reasons, the first is the hassle at boarders and the legal restrictions; the second is far more serious.

If you have a rifle and you meet another person with a rifle; and they decide they want your rifle or that you are a threat... it becomes kill or be killed. If a man raises and aims a rifle at me and I am armed, I will either shoot him or be shot far more likely than if I were unarmed. Unarmed I am no great threat, and he can just steal my stuff. Armed, the dynamics change. In a place like America though, where firearms are far more accepted, I would definitely carry a rifle in accordance with state and federal regulations to hunt along the way.

But I do carry a tool to hunt with. I carry a light weight and very powerful Mongolian horse bow similar to the ones used by the Mongol hoards that swept across the steppe hundreds of years before our ride. It's great fun, and can be used to harvest small game and can easily shoot a fish mooching along in shallow water. It's a bit of a hassle to carry with us, but it's brought us a great deal of enjoyment and has been a real novelty with folks we meet along the way. It's powerful enough to kill a small deer, jackal, fox or rabbit easily at close range, and has put a bit of meat on the fire for us, though nowhere near as much as I had imagined it would.

Mainly because I just don't feel the urge to hunt all that much. With a full belly and a gorgeous woman by my side, the thought of sharing an afternoon by the campfire with her and the dog is more appealing than stalking alone through a forest where I am most likely going to be poaching game indigenous people rely on for winter survival, or is legally protected. Still, knowing I have a bow by my side as I listen to the wolves howl in the nearby hills... I have to admit; it gives me piece of mind.

Fishing has been one of my favorite pastimes since sometime around my 10th birthday. I've fished the world over, and I have done a bit of fishing during the ride; but never as much as I had hoped I would. Much like the hunting, we've found it pretty easy to keep food in our bellies and in many places the fishing is legally restricted. But carrying a small telescopic fishing rod rolled up in the tent roll as well as a small amount of fishing tackle has been awesome fun. I've got a few fish along the way, with a few real monster Taimen "the worlds largest type of trout' in Mongolia being the highlight!

A few lures, a few small hooks and sinkers as well as a rod and light reel take up very little room and you'd be surprised just how big a fish you can get on light gear such as this. Fish of course can easily be bought along the way too, but there's something romantic about the idea of picking a bit of wild thyme and sage and cooking a trout you've just plucked from a stream on a rock by the river side over a camp fire. It's for this reason I will always carry a rod, and bow. The romantic idea that I can put a meal on a fire and feed us in a wild place, with little more than the tools people have used for thousands of years.

When it comes to foraging food in the wild, once again smart phones have revolutionized the way it can be done by the tec savvy traveller. During our time accompanying our friend John, the restaurant owner and wine grower in the mountains of the Tusheti region of Georgia; we spent a lot of time foraging in the forest. He was mad keen on mushrooms, and although I don't mind normal old field mushrooms fried up in butter; the idea of eating weird looking fungi doesn't overly appeal to me and it is with a very reserved caution I will just start eating mushrooms without first an expert's opinion.

Mushrooms; you can find them everywhere and although there are plenty that can make you very sick, there's a ton that taste great. Did you know for instance that the underside of a mushroom will have either fins, or look like a sponge? If it looks like a sponge, your good to go as there is only one type of mushroom that has a sponge underside that is poisonous. So... as long as you know what that one looks like, everything else is good to go; and with an app and an appetite you can sort yourself a good feed that could easily be worth a fortune in a high-end restaurant.

Porcine mushrooms are a delicacy I'd just walk past, and saffron milk caps literally taste like chicken! Let me be clear; if in doubt, don't eat it. Use the many resources available to ensure you are well educated before you go looking for food, and have the ability to check that what you think is food; is actually safe to eat. Well, there's an app downloadable for it; hell, there's an app for everything and there are many apps available to help you easily identify edible mushrooms, berries and plants in the forests of Europe and Asia.

I honestly had a great time foraging with John; and learned a ton about wild food, mushrooms and herbs in particular. I'd always picked and eaten blueberries, strawberries, blackberries and raspberries as all are easy to identify, find, and taste great. But finding herbs was a new one for me, and I was amazed how easy it was to find a huge variety of herbs that could be used to season food as well as make amazing wild teas. I wasted 41 years not drinking a cup of stinging nettles with black thyme, a pinch of mint and some St John's wart; a true gift from the mountains.

I hated stinging nettle; cursed the dam stuff whenever I saw it or had it brush against the back of my hands or legs. But as a tea? In soup its amazing too! After a bit of time in the forest with a knowledgeable forager like John, the way I viewed the foraging in the forest changed overnight and forever.

Hunting, fishing and foraging can be a way to supplement living in the wild; but trading and purchasing goods from the locals is a far more realistic way of surviving on the steppe.

Chapter 11.
Reality check.

"A man with outward courage dares to die; a man with inner courage dares to live."
Lao Tzu, Tao Te Ching

We were short on water as Luisa really put it away over the last 15km or so of empty steppe, loosely following a road towards what appeared to be a reasonable sized Kazakh town. I was sipping away on my Vodka as usual, as Bill the Bastard plodded along methodically behind Luisa on Jac. Lu was chatting away to Jill who as usual was happily darting about under the horse's feet and where ever possible staring lovingly up at Lu from her heel. She would do this all day when she had energy, but when she got a little worn out she would tuck in behind the horse and trot along with her face just below the horses tail.

There were so many times Jill would just sort of "get in the zone" so to speak and not notice the horses tail lift. But I did... and a grin would creep across my face that turned to full blown laughter as a steaming pile of horse poo dropped straight onto Jill's face knocking her out of her little doggie trance.

We often wouldn't talk, though neither of us ever really shut up when face to face. Sometimes like this when we were a bit tired, the horses felt a bit sluggish and supplies were low we just sort of "looked out the school bus window and planned our homework". There was no standing grass, and hadn't been any for a long while and as we drew closer to town; it appeared deserted, and I do mean really deserted. The usual wandering stock were absent as we passed through the rubbish dumped everywhere on the outskirts of the soviet era village, painted in the standard drab, faded shades of pastel and headed towards silent empty streets.

Something was wrong; you could just feel it.

Chickens were about all we saw for the first few blocks, and even the stray Kazakh hounds were nowhere to be seen. Lu and I looked for a, well, a magazine "Kazakh for shop" or anywhere we could draw water and find a few beers for me, when a giant hobbled stallion charged at me and Bill, almost causing me to drop my hip flask. As usual, the graceful weapon of a woman on her gentle red gelding charged in dominating the challenger leaving me to manage my fizzing stallion as we pressed in towards the heart of the town.

Finding a main street of sorts, we noticed a mosque some 500 meters further down the deserted streets surround by cars, and when I say cars, I mean most likely every car from every village within hundreds of kilometers. As we drew closer still, it became clear it was a funeral and two filthy strangers on horseback looking for hay, water and more booze may be bringing disrespect with them.

We ambled towards them, looking for a way around the mosque in order to pay our respects to the departed when a black mid-eighties BWM approached at a great rate of knots. They pulled up next to us, and as usual we were greeted in Kazakh by confused, but welcoming faces. At first, we couldn't hear them over their pumping Russian house music, but we did manage to understand the word "Akuda", which as best we could figure means where are you from? Like I said, I'd had a few vodkas and began saying

"Akuda Straya mate"

as I did my best impression of a kangaroo. Lu laughed, so did the Kazak's and as Lu said

"Akuda Germania"

I held back the urge to make some sort of inappropriate WWII reference... yes, I make inappropriate jokes far too often given our grandfathers, and great great grandfathers served our respective nations of birth.

The drivers face changed on hearing "Germania", and he suddenly broke into banter with Lu in her native tongue as the rest of the car chuckled away at my kangaroo impersonation.

It turns out it was a funeral, and that it would be best if we gave it a respectfully wide berth which we did; but the town had an English school teacher living nearby, and that they would send her son to meet us, take us to a shop, and then help us with anything we needed from there. Lu chatted away in German, as I slipped off Billy and stroked his long gentle nose and stared into his beautiful soft, kind and soulful eyes.

I slipped his saddle off, dumping it crudely under some poplar trees at the edge of a children's playground. Most of the leaves had fallen, and as Billy shook himself and sighed like a soldier removing his body armour at the end of a patrol, he began to graze among the fallen leaves. I did the same, leaning against my saddle and tucking into the beetroot and cous cous salad the "Vegan chef" whipped up the night before. As Luisa chatted away, I lay back, staring up at an endless blue sky obscured by the last yellow leaves defiantly clinging to what remained of their summer home.

The BMW was soon gone, and the silence of the steppe was only broken by the sombre and sincere prays we could hear coming from the nearby mosque. As I lay there staring into space in a warm vodka induced haze, I started to think about this amazing host nation, and how coming here without having researched what to expect was the right way to do it. Think what you want about Muslims, think what you want about Kazakhstan, but believe me when I say your most likely not even close to the reality of life here. You're wrong about everything you think you know if your opinion is based off the mainstream media and movies "Borat".

These people are beautiful, and I have seen first-hand the side of Muslim faith that truly does deserve to be known as "the religion of peace". Borat didn't even get it right about Kazakhstan being the number one producer of potassium. I snapped back out of my day dream as a young boy in a yellow t shirt sped up on a BMX, locking up the back brake just short of me and throwing the bike sideways; stopping just centimeters from me. He looked at me in silence for a second with and enquiring glance before a huge smile broke across his face and he said

"Kangrooo, kangarrooo!"

I was soon at the trot beside Lu, with the little dog staring up at dog mum with refreshed excitement as we followed our new little mate down back alleys and streets of his home town. He swung his bike from one side of the road to the other pointing out land marks, and points of interest in the few English words he knew.

"School, school!". "Magazine shop, close close, tomorrow, open open". "Home home, Mami, Dadi, sister sister sister, brother!"

He was so excited to show us his home and we were soon being welcomed into a beautifully well kept courtyard with huge stores of winter hay and a beautiful spring fed trough of crystal clear water the boy excitedly showed us.

"Voda voda, loshad loshad voda!"

The boy excitedly led our boys to the "water for horses" as his mother, the local English teacher and her husband excitedly joined us, rushing over to enthusiastically shake our hands, hug us and welcome us into their home.

Lu and I both paused, focused as always first and foremost on our horses only to see a young woman grinning from ear to ear bringing them hay as our little mate with the bike stripped their saddles and tack, placing it neatly and with great care at the rear of their home. Vodka. Much vodka followed. And the warmth of it was exceeded only by the love shown to us by these beautiful people.

We were fed, as were our horses and offered every kindness imaginable.

The father was a shepherd, and he looked with great admiration at my Swazi micro-fleece t shirt, so I took it off and gave it to him. Stunned, he took off his hat and gave it to me. The ladies chatted away and our little mate with the bike showed me endless piles of trophies earned in the boxing ring. In days to come, I sincerely hope to rise to my feet with tears in my eyes screaming at the tv as he wins "by knock out" in the first round of his UFC debut. It was still early in the day, and Lu and I knew we must press on despite the offer to stay as long as we wanted.

As much as we love the endless hospitality shown to us by these beautiful people, Luisa and I enjoy the steppe and the sense of solitude and peace we feel alone with our horses as the sun sets. With a full hay bag, water, food and vodka I stood by Billy and hugged... no. Held a brother from another mother very drunk in a Kazakh courtyard of a town I don't even know the name of. We felt a closeness. Respect for one another and a bond, unspoken; but felt.

He was a horseman, a true wild horseman and shepherd of the steppe, forged in a land that would soon see temperatures of -40 for months on end. We were two fools that shunned comfort and warmth for the steppe, and he loved that. He respected it, and as I mounted my stallion he nodded in approval and proudly tugged on the sleeve of the shirt I'd given him and smiled. As Jac turned anxiously in anticipation, and Billy again began to fizz beneath me. We shared one last look, and parted with the words...

"Peace be with you"

Bill the Bastard burst into life beneath me, and the true degree of my day drinking became apparent as I dropped the reins, closed my eyes, and tilted my head upwards towards the sky urging my stallion to the gallop. This is freedom. This is what it means to live without fear. This is how you suck it all in, chew it all up and feel closeness to... God? A greater power? Yourself? Or is it just love of life? Is it acceptance? Or am I just some drunk fool laughing beside a gorgeous "and I do mean gorgeous in every sense of the word" woman as we gallop through the straw yellow Autumn grasses of the Kazakh steppe as the sun begins to fall?

All I know is this is a feeling I cannot word easily as it is just that. It's a feeling. One worth searching and fighting for. One worth living for.

We rode on some dozen kilometers or so, and soon; we were completely alone. We paused, began to drop our gear for the nights camp when over 100 wild steppe horses appeared. Bill and I joined them, and galloping among them on my wild Kazakh stallion; I made my peace with many of my demons... and those that remained would face me on horseback.

This is who I am; I am accepted here, and I belong. I was falling in love with Luisa, and Bill the Bastard and I had become one. I had a horse that I could ride into hell, a girl that would be there riding just as hard by my side and Jill.... Getting poo dropped on her head following her dog mum like a shadow on a cloudless day.

Righto... so that kind of came out of nowhere and wasn't even what I'd intended to write, but I think in regards to this chapter it's perfect. What you take from it is up to you, but for me it's simple. You may have no idea what you're looking for; hell... I know I had no idea at all and that's the beauty of a journey like this. You need to be open to the idea it's never going to be how you plan, or what you think you want, but may end up being everything you need without knowing it.

Budget for gearing up and getting out there.

You don't need to sell your house, you don't need to inherit a fortune, and you don't need to win lotto; you just need to be realistic about what you've got and what you wish to achieve.

You can live very cheaply throughout Asia and live to a high standard too. Guest houses, Air B n B's and even the hotels are relatively inexpensive when and if you choose to stay in them. It's a good idea to take a break whenever you can, shower, charge up any electrical gear you've got and get a good rest in a bed.

You will of course be offered an endless stream of free accommodation along the way too, but getting a good shower there is rare, and beds are usually just a rolled out mat on the floor. Staying with locals is great, as they will often give your horses a really good feed and you can sleep easy knowing they will be safe from wolves or horse thieves. Accommodation starts from as little as $3 or $4 USD, and we've stayed in some really nice places for around $15USD a night.

Our COVID-19 forced pause that we are in part using to write this book is costing us a $150USD for a full month in an inner-city apartment in the Georgian capital Tbilisi.

Food and drink are dirt cheap everywhere you go, especially if you cook your own with fresh organic vegetables and milk at prices you literally can't believe if your used to purchasing these similar items at home in the west. I can remember paying $10USD a kg for tomatoes out of season at home in NZ, and the last lot I bought here were around 10c USD a kg. Fresh milk in summer here is as little as 1c USD in the mountains... and in the shops in cities its rarely more than 25c USD. Beers cost bugger all, usually about $1usd per litre and wine, vodka and anything other than imported top shelf can be just as cheap.

A meal in a restaurant varies, but this afternoon we are going out for a $9USD deal that includes a pepperoni pizza and four beers. How you eat, where you eat and what you eat is up to you; the same as where and how you choose to spend your nights, and realistically Luisa, Jill and I both live very well on a budget of around $100USD a week between us all on the steppe. With a budget of $10,000USD behind each of you, I would consider a pair of like-minded travelers potentially able to live much as we do for up too two years!

The budget needed for your gear isn't an easy one to put a figure on for a number of reasons, and the following is at best a "rough idea" of what you can and should potentially expect to spend. It's up to you to decide the quality of the gear you wish to buy before you set off, whether you buy brand new or second hand, and if already traveling just decide to buy a cheap local saddle and go for it!

As a career outdoorsman, I have a significant amount of money invested in gear, fortunately much of which I share with Luisa and had my spare gear sent over for. I do not for one second regret having invested around $1,000USD some years ago in a top quality MacPac tent, the same type you will quiet likely find at Everest base camp.

Our Swazi gear isn't cheap either, nor my sleeping bag or boots; it's these items that all are the difference between life and death as well as just comfortable living outdoors.

Between Luisa and I it is fair to say we have around $5,000USD worth of gear, and much of it is second hand or acquired cheaply along the way. If you looked for bargains, begged borrowed and stole bits and bobs from family and friends I think you could be adequately kitted out for less than $1000USD, but just how long your gear will last is a reflection on the quality you've paid for. Good is rarely cheap, and cheap is rarely good.

My saddle is as best I can tell worth well over $1000USD; but I picked it up for $100USD give or take second hand several years ago. It is an "Australian Outrider" Aussie stock saddle with a wooden tree and horn that is very solidly attached. Being an older but very well-made saddle it's been utterly brilliant, and other than the high cost of a quality saddle, much of the other gear required for a horse is very cheap.

I feel confident I could walk from the black markets in Ulaanbaatar fully kitted out to ride off into the sunset with change from $200USD if you were wanting to just make do. But I do swear by my soft French snaffle bits and mohair girths of course, and even these are relatively cheap although impossible to find overseas. Between Luisa and I, I'd say we have spent less than $500USD kitting ourselves out, but if we were at home planning a trip now, I'd buy us both matching stock saddles, and build two identical long ride setups that I feel pretty confident would set us back around $750USD each, made with all brand new items and be perfectly suited too years of daily abuse in the wilderness.

Purchasing a horse.

I knew bugger all about picking a horse, but I could do a half reasonable job spotting a lame one and so could Luisa. There are a ton of articles online, as well as a heap of groups on Facebook where they talk endlessly about a horse's confirmation. But that's not what I look for at all. I look into its soul and listen to its fears and needs and see if he's going to be a mate, total nightmare or just a companion.

I'm after a mate; and a mate that will keep me safe and enjoy his time with me as much as I will with him. I've found this in Digger, Bill the Bastard, Pippa and Gorda; I hope to find it in many more horses I work with in the years to come.

I had it with Beersheba too, but he just had too many vices, and we didn't have enough time to ever get him right. A horse's eyes, and the way it responds to your touch are indicators that mean the most to me, though athleticism and spirt are also huge factors effecting my selection of a horse. Temperament. Temperament is what makes or breaks a long ride, as you just can't spend your entire time fighting your mate, and for this reason above all else, it's my number one selecting criteria.

The breed of horses available will change with the scenery, and a good sturdy little horse on the steppe is a petty sound bet if your buying in central Asia. Check they have good feet, if the horse will allow you feel around inside its mouth a bit too, make sure there's no sharp bits of tooth irritating the side of the horse's mouth and look for one preferably of no less than 7 years in age and as good a condition score as possible. If possible, try and get a stock horse as they will have been loaded on and off vehicles usually, and are far less likely to spook. But be aware you may be limited in your options, and have to chose a horse that is "good enough".

Have a measure up using the formula from the chapter 9 and make sure the horse is able to comfortably carry the weight you're asking of it. Check the saddle fit too of course, and take it for a ride feeling for any hints of lameness and if at all possible, push the horse to do a few things he doesn't want to do to test out his temperament. Grab out a beer and crack it near the horse, shake out a tent, and see how readily it will allow you to catch it and handle it. Be VERY careful doing all these things, as these are often pretty wild horses on offer and although they will potentially be your best mate in a week's time, right now they need to be gently tested and felt out.

Mares are always going to be a bit "mare like" as they are the bosses of the wild herds and paddocks at home. I don't ride mares, and for the most part Mongols never ride them. On the wide-open expanses of the steppe there are tons of wild horses and if your mare comes in season, there's no two ways about it; she's going to end up in foal. I'm told most great endurance horse races are won by mares, and may be perfectly suitable for your needs in Europe or another developed part of the world. But for me personally, mares are out leaving just stallions and geldings as your two real options.

To be honest, it's rare for any real issues to be encounter with a well ridden gelding used for stock work, and if you're a beginner rider, this would be what I'd recommend you aim to purchase without a hint of hesitation.

I was always very reserved around stallions; and had been very poorly advised by my ex in what I have now come to realize was a topic of which she knew less than nothing about. Stallions are just horses, and horses need boundaries set for them, and maintained by their rider. If you let a stallion take liberties, he will and you will end up fighting a testosterone filled horse weighing half a ton, and that's no fun at all. But if you set the rules confidently and firmly right from the start, then a stallion can be an absolute lamb.

He may look at a mare and think... might go get me some of that. But if he looks at a mare and goes... I know this game and I better not do anything stupid here. The boss will give me a bloody good whack if I try and buck him off and head over; a nice soft pat if I don't, a good brush tonight and some carrots and hay... so yeah, I think I'll be a good boy. This goes for any horse, and although a gelding is generally a lot calmer, my stallions have all been so calm with me you'd never even guess they were stallions.

What you can expect to pay is going to vary a lot too, but the general rule of thumb is they will try and charge you whatever they think they can get away with charging you. Horses worth purchasing start at around $400USD, and when you on sell your horse expect to lose about a third of what you paid at least. It's just how it goes sadly, tourists pay tourist prices, and you have got to sort of suck it up and pay what you need to get what you need. When you do pay, make sure you get a photo on your phone of you handshaking its old owner and handing over the cash with the horse in frame.

Also make sure and get a receipt and any ownership, registration or veterinary paperwork you possibly can in case you are ever questioned on the legal ownership of the horse. If you can get a native speaker to help you purchasing, it's going to save you a lot of grief and it's a very small investment to shout them lunch, the taxi fare out to look at horses as well as putting a few bills in their hand at the end of the day, as it will save you a fortune.

One last point worth mentioning; I'm a big believer in purchasing the most plain, dull looking horse you can find. Pretty horses get both stolen more often, and often cost a bit more just because they are pretty. But if you find the right horse, regardless of color, breed or price, it's worth the investment.

Timing your ride.

Luisa and I crossed Mongolia in a little Toyota NOT on horses. We did this because winter was coming and her visa had mere days remaining on it. We have both had our own horseback adventures in Mongolia individually, and for things to work despite my dreams of just riding off into the sunset from the north of Mongolia; reality dictated otherwise, and I arrived as the last of the Summer started to fade into Autumn, and Winter on the steppe is cruel; like... one mistake and your dead cruel.

With temperatures that regularly hit -40, if you need to go for a poo it will freeze solid to you on the way out if you take longer than 30 seconds. Think about that for a second... not to be gross. But because it's the kind of thing I for one would never have thought of before going to Mongolia. Think about what happens when a kid licks a frozen ski lift and his tongue sticks, needing hot water to get his tongue free... Yeah, that can happen to you doing something as simple as going to the toilet! If you get things wrong, you can die.

Winter on the steppe isn't your friend, but you can survive it, and a New Zealand based adventurer even runs tours across the Gobi Desert on camels during winter! But for the average adventurer, us included, we decided that -20 was as cold as we were comfortable operating in and packed and planned to survive this comfortably.

But timing when you go around the seasons is not the only consideration regarding the timing of your ride. It may take you a day to find a horse; but maybe not. Maybe it takes 2 fruitless weeks of searching and you end up 300km from where you thought you'd have started from. The point I've meandered towards is this; flexibility. I would think saying to yourself; I'm going to go, buy a horse by day three, ride for two weeks and then fly home by day twenty may not be completely realistic.

If time is something you are short on, then I would defiantly recommend booking one of the endless number of horse trekking holidays that are available the world over, and in the acknowledgments we recommend a few people who can help with that. To be honest, it may even end up being a better experience with the language barrier taken care of and horses that are used to tourists trained and geared up ready to go.

BUT... be aware many of these sorts of operations are "in my opinion" pretty good at taking advantage of locals, charging western prices while paying local wages.

If, however, you've decided that you're going on an adventure; by that I mean a real adventure! One where you plan to just get on a plane and go, see where the wind blows you looking for love, new experiences and that cliche line "finding yourself", then you need to be flexible. For me, flexible means cutting all your ties, chucking anything you want to keep in a mate's garage and walking out the door with no idea when you're coming back, if at all. That is how I timed my trip, that is how Luisa timed hers, and I tell you from the heart; it feels amazing. But there are constraints, and they are visas and borders.

Visas and borders

In the post COVID-19 world I hold the worlds 2 strongest passports, being New Zealand and Australia. Germany is rated the safest country in the world for COVID-19 and Luisa's Germany passport is just as strong in many regards.

Having a strong passport helps no end when it comes to travel, but there are some things you just can't get around sadly, and that is the lengthy visa application required for travel through Russia. Luisa and I got "transit visas" in a matter of days through the Russian embassy in Ulaanbaatar. It cost me around $250USD to cross Russia to Kazakhstan where we were just welcomed across the border with a smile and a bunch of patriotic AK47 carrying border guards shaking our hands, excited to see western tourists with saddles!

All boarders differ, and visa requirements are something you need to look into prior to even beginning planning routes or purchasing gear and the strength of your passport will affect where and how easily you can travel. Be aware that if you travel through Iran, Syria or any nation the US has current beef with, you will have a lot of questions to answer on your return to the States if you're a citizen. If you're on a foreign passport trying to get into the US after having spent 3 months riding horses through Iran, again you're going to have some explaining to do and will no longer be eligible for the visa waiver offered to many countries. You will need to apply via a US embassy for a full visa which can take months.

There are tricks for getting visas, and people you can pay to do the paper work for you, but I have found the best way to get things done is approach the nations embassy directly and tell them exactly what you want to do. China requires a detailed itinerary of your travel plans, as does Russia, and there was just no way I could get through there on horseback being honest about what I wanted to do, without spending months negotiating it.

Be aware that in countries such as Uzbekistan, although welcoming and safe, you are required to show upon exit that you were registered in hotels along the way. You can move about, but are required to either change region each week or have registered hotel documentation to show if questioned by authorities.

We messed it up a bit in Uzbekistan and questions were raised; we were almost deported as we had technically broken the rules, but as we had made every effort to do the right thing beforehand including having met with the regional minister for tourism, we were allowed to stay. It gave us a bit of a fright I'll admit, but they were great about it all, and the police and immigration were very understanding, and when the situation was explained to them, they were just in awe of what we were trying to achieve.

When questioned by police, producing the business card of the tourism minster certainly made the difference, and I would defiantly recommend you let authorities know exactly what you're up to and where ever possible get some sort of letter or documentation, even if it's just a phone number for a situation where a confused authority figure may have questions you can't answer.

But our friend Gunter, whom we travelled through Mongolia with, had applied for a Russian "business visa" before he left Germany. It gave him multiple entries and a full three months free to travel unrestricted through the Russian Federation. These need to be applied for before you leave your home nation, but could be a viable option for someone wishing to ride through Russia, but still: this doesn't overcome the issue of crossing a border with horses.

No matter how many times we have been told that "it can be done", we have struggled. We have struggled a great deal because it all takes time and money we don't have. If you invest years in planning, send endless emails and have great contacts in country, then I have no doubt it can be done. But for us, we just made the most of what we could do easily and that was get visa's, buy horses, ride as much as we could and then sell and cross to the next country with as little difficulty as possible. Sure; the dream of an uninterrupted horseback journey is always the ultimate goal, but in practice it's very difficult, and we feel the way we did things allowed us to get the most out of our time with the minimum hassle at boarders.

Do the research prior, email the embassies and DON'T LIE. You can leave the odd thing out of an application and pull the old "I didn't know I needed to include that", but if you lie on a visa application then you're not getting in and there's every chance if your there, you may be staying longer than you planned under less that desirable conditions.

Bringing a dog.

In short, it can be done, but it is going to cost you in both time, and money at every step of your journey.

With a "pet passport" you can obtain from a veterinarian in almost any country and up to date vaccinations specifically a rabies antibody test, it is actually very easy to cross borders with a dog. But Jill costs as much as both of us put together in added expense at times. If you catch a train, you need to book a cabin not a seat, and often travel requires a dog travels in a pet box of sorts. She needs vet checks, shots and although we can just swap out the horses at the boarders where needed, Jill is here for the duration.

Jill is also a constant target for aggressive dogs and very good at putting herself constantly into harm's way. She's a bloody endless hassle to be honest, but we love her. She's just so loyal; even if she makes it hard to catch a taxi at times, and halves the number of places we can get accommodation it's all worth it to see that smiling little face bouncing alongside us all day long. Her cute little mannerisms as she begs for food, or convince us her spot on a cold night is right between us, nuzzling her way under the sleeping bags or just gentle head on the lap by the fire.

Bringing a dog is a risk and expense, but also a huge reward that both of us feel is well worth the effort. But there are days when I curse her and want to never see her again. In the case of a dog attack for instance, two riders simply gallop away. But with a dog, you have to dismount and fight, ride with her in your arms or abandon the dog to its fate. No matter what option you chose, you are all put at considerable risk because of the dog that could have easily been avoided by her being left at home

Things break, you break, and you better know how to fix them.

I learned to repair my gear as both a soldier and as a hunter. A lot of gear just isn't up to the task it's issued for sadly and the Army of almost every nation has its troops make do far too often. So; we learn how to "acquire" the things we need from other units or armies and often modify it to suit our needs, and often paint it in such a way it no longer resembles its often questionable origin.

Serving with the New Zealand Defence force as a machine gunner, I wore webbing I had traded with an Australian SAS trooper, and had simply painted it so it didn't look out of place. It allowed me to carry 1200 rounds for my machine gun, a pistol, 2 grenades and 2 smoke grenades rather than the NZDF issue of the time that allowed me around 600 rounds plus the grenades. I learned to sew with a sewing awl, constantly modifying, padding and patching the webbing I carried allowing me to carry more and more in comfort, jump in and out of vehicles without issue, and work as effectively as possibly.

With my saddle bags, girths, and even the gear I ride in you will see patches, odd bits of fur and strange additions picked up along my journey. With no sponsors and limited finances, all our gear is hand sew and modified by us, to fit us. This constant evolution has allowed us to develop our own style, and one we are proud of, having failed so many times with gear we were told would work.

I mentioned a good horseman always having bail twine and a pocket knife; well, knowing what to do with them helps, and the only way to learn is to get out there and fail with the right mindset. When something breaks, it's an opportunity to upgrade the part so your less likely to fail again. You can't just buy replacement parts in the wilderness, and you can expect even the best gear to have a breaking point. Further to this, the stuff you have will always have a few areas it can be improved, and both Luisa and I have both said...

"Yep, I'm finally happy with my set up."

Only to pull it apart again a few days later. Be prepared to constantly repair and evolve; furthermore, you'll enjoy every minute of it.

Reality check

Luisa and I both weigh 65kg and we work very hard to stay both a suitable weight for our horse and suitably saddle fit. It's not an easy task by any stretch to keep weight off at 42; not like it used to be anyway. We eat good food, often only two proper meals a day and despite my drinking too much we keep hydrated and walk Jill any day we don't ride with her. When were in town for whatever reason, we walk everywhere, and as any rider that has had a fall and spent a few weeks out of the saddle will tell you, saddle fitness is something that can be lost very quickly.

Even if you ride 2 hours a day every other day; that first full day in the saddle is going to hurt, and it's going to hurt a lot more if you're not prepared for it. But by the end of the first week of a long ride it all just falls into place, or falls apart. If you're going to have any serious issues with your kit, horse and your body they should show themselves pretty quickly; and for us its normally the third day we've had our worst failures.

Go into an adventure like this knowing it's going to be hard. You're going to hurt, you're going to need to constantly grow your abilities and know both your strengths and weaknesses while checking your ego. If you don't check your ego, a horse will check it for you. Don't ask for or expect an easy ride, easy horses and gear that never fails. Ask in yourself for the strength to push past all and any obstacles you may face or for that matter... fear.

Chapter 12.
Letting go.

"Buy the ticket, take the ride"
Hunter S Thompson.

At home it's so easy to hide in the crowd, talking about the trips we may one day take, about our hopes and dreams from the security of familiarity. It's not until I return to the cities, where we humans live like a schooling fish, all struggling to hide in the middle of the school so as not to face all those scary things outside the safe boiling mass. All the fish are the same size, and all try and swim in the same direction, just like society. And just like fish, we grow to fit the size of our surroundings. A small lake with only 50 perch in it will have large healthy fish, but the same water holding 5000, will have tiny perch, rarely growing to more than 1/10 their potential.

Lu and I both chose not to shrink and conform to society's expectations; we broke free, and by some miracle, found each other swimming alone.

Jill.

The noise of the blowflies around my ears had me shaking my head again and again, maddening really, but what can you do? Red dust sticks in my eyes, my mouth, and covers my shirt in a marble like randomness in utter contrast to its bright green color. For only a second, I take my eyes from the limping calf at the rear of the mob, grabbing my water bottle, only to be reminded I'd emptied it more than an hour ago.

My eyes rise to the blue sky above, knowing the blue will soon fade to black; I take up the reins and press on, hurrying to get the beefies in camp before losing the light.

I enjoyed stepping up the tempo after another, long, slow day in the saddle. I would have never thought how boring droving would be; I knew nothing when I decided to pack my bags just after I finished school to escape for a gap year to Australia. The promise of a wild country still full of cowboys, horses, dogs, and adventures; I found all these and more, but none quite as I expected.

My thoughts would wonder often, day dreaming, but always loving every second just sitting there on my horse, who to be fair, did the lion's share of the work for us during the day. But this evening I was in full control, pushing and working the beefies towards our nights camp. After a long monotonous day, I was into it until my attention broken by the lumpy rumble of an approaching quad bike, diverting my focus from the brahmans I'm pressing forward.

I can only smile as a stranger approaches; a weather-beaten face buried under a cowboy hat, a ripped flannel shirt and the dirty jeans telling the story of a long day's work in the cattle yards without the need for words. As he stops and rolls a smoke, the boss appears on his bike and the two stop and chat together as I press hard on the lame calf, ensuring it mobs up with the rest of the herd before darkness falls. It does, disappearing among its kind leaving me free to join the pair of stockmen for a chat.

I already knew what their conversation would be about... the weather, and to this day it still mystifies me as to how farmers in a dusty, dry, sun drenched land, can spend an hour talking about rain. As I trot towards the stockmen, I think back to the first conversation I had with farmers about the weather, I spent most of the time just nodding and looking into the sky.

Yep, I guessed right; as I arrived, I was welcomed with a nod, and they return to their conversation about the chance of coming rain.

The boss breaks off briefly from his conversation, and we have a quick chat about the days muster, in particular the lame calf. He assures me the camp is just around the corner and that I should push on, press up hard on the mob as the others were already at the yards as it was dam near beer O'clock. I nod in acknowledgment, taking up my reins and dream of that first cold can; but after turning my horse to trot off, the stranger fires a lazy question my way...

"Hey, ya wanna a dog? Jill's er name, she's a youngen', no bloody good in the yards and that's where I do me days, not out here droven like you fellas. She could be half good at that in the right hands... whatcha reckon, Kraut shelia like you and her could be a good match if ya play ya cards right... whatcha recken?"

I turn my attention towards my boss; he only looked up for a second as he lit a smoke, shrugged his shoulders and said...

"One more dog won't make any bloody difference, and if she won't work half decent we'll just bloody shoot her yeah..."

The stranger, like me, didn't seem too impressed by his last words, and again looks too me with a growing intensity as if he were trying to say more than he could ever put in words, but he did find a few to try and hammer home the hard sell of a useless dog he clearly held no ill feelings towards...

"She's real loyal... be shame if she never got a chance to find herself a partner"

My eyes start to sparkle, and that crooked smile began to spread across my filthy face; biting my lip in excitement... tingling at the thought of what I am just about to do, I answer.

"Yeah... thanks."

The conversation was over in seconds, and I turned back to the mob and gave my horse a stir as I smiled, knowing a dog I'd never met would be waiting for me at the yards this evening. A colorful Australian Koolie x Kelpie bitch with the most beautiful eyes, one blue, and one brown, was waiting for me at camp that night. She was in back of the stranger's four-wheel drive on a short chain, I saw here heading into camp before even getting off the horse and immediately felt tingles all through me. Soon after, we had our first walk together, taking my new, timid little mate down to the water for a drink.

From here, magic happened. As my training with her began, every day we drew more and more laughter from my workmates. I saw myself when I looked at her, I knew how misplaced she must have felt when they tried to make her something she wasn't, undermining all of her beauty her unique nature. She is a disaster, a terrible working dog, but is that all that counts? It made me want to scream some days; why can't anyone see how beautiful, how loyal she is? But just like me growing up she was out of place, and when we found each other... we would both be never be out of place again as long as we have each other.

I took her on without thinking about the consequences, with no real plan as such, thinking I would keep her as my dog for now, maybe passing her onto another backpacker before returning to Germany. For just one summer I wanted it to be real, a connection, magic, I'm not sure; just something perfect even if it were just for a moment, and in what seemed a blink of an eye, that moment passed. Three thousand Euros... I can't believe I'm considering this; taking Jill to Germany.

When I went to bed that night, I tried to talk myself into leaving her here. Maybe I could find a nice home for her? I felt something like relief for the briefest second, but her little face appears in my head. The love she looks at me with daily just left me hating myself for even thinking this way. Rolling over from side to side, I finally escaped into the world of dreams but reality hit home again the following evening.

I splashed cold water in my face, already smelling dinner while washing my hands; they are so little, square, worn from work and not at all lady like. I look in the mirror, not liking one bit the woman staring back at me; I turn away in disgust. I can't believe I'm thinking of leaving Jill possibly, forever. I still had about 30min before we would sit down for dinner, so I grab a beer out of the fridge and disappear into the backyard to be alone and think.

Autumn colors great me, everything around me fading as winter approaches. The red and brown leaves, a shadow of their former selves have my imagination wander to summer days full of vivid greens, colorful flowers and time spent enjoying companionship with Jill.

It seemed like a warning, that maybe everything has to come to an end someday, right? Cold beer runs down my throat, and suddenly something changes inside me. Love, it doesn't change; the seasons do. My body starts to tingle, just as it had the first time I saw my little girl, feelings growing inside me; a lioness inside me had woken from long sleep, hungry, determined and ready to protect her cub at all costs. I quietly whispered to myself...

"Let's get you Home Jill"

But I did leave; briefly, to renew my visa and I returned within a only few months, and when the time came to leave Australia once more, I had spent a full year preparing to take Jill with me. I knew it was only one path to take; out of respect for myself, the love for a little dog and my faith in a magical life whose existence I was still unsure of, but had glimpsed, and was going to keep searching for. Despite every obstacle, we made it; and the feeling when I opened her cage at the Munich airport... it made everything worthwhile. Just as my decision to take Jill home with me to Germany was led by my heart.

I took another chance 4 years later, choosing to ride off into the sunset with a total stranger. But now that girl once too afraid to even go to sleep, lives her dreams, and although seasons change, my feelings for my partners won't. The beauty we've shared will last a lifetime.

I'm not crazy.

We'd had a long day, and I was worn out watching the sun sink over the empty grassland of the rolling steppe; now bathed in a peach orange glow as the sun made its final visible moments as dazzling a display as possible. I was standing less than 20 meters from Luisa as she cooked us dinner, with the horses grazing behind her and Jill sitting beside her watching dinner cook; as usual, begging for food. We we're in a small depression in the endless steppe some 15km from the nearest other human, soaking in the serenity of this amazing expanse of literally nothing. I was pondering the reality that in less than 6 weeks, everything here would be frozen, and hidden under a meter of snow when I saw something not of this world.

A mercury silver metallic disc, approximately 15 meters wide and 2 meters high shot straight over both horses and then Luisa at a height of no more than 25 meters, making a slight whooshing sound, no louder than a large bird in flight. It was gone as fast as it had appeared, and didn't even alarm the horses, but it was enough noise for Luisa to stop cooking, look up, and ask what type of bird made the noise?

I just stood there holding my beer with a stupid look on my face for a minute or so before I spoke. I was trying to figure out the physics of it all... I mean it was going so dam fast it must have been breaking the sound barrier; but there was no sonic boom. Did I imagine it? Was it some trick of steppe, played by a reflection off a nearby lake as the last of the days light slowly faded through a spectrum of Autumn hues in this empty wilderness? Or was that some sort of straight up Bob Lazar on Joe Rogan's Podcast area 51 alien technology. Or was it Russian?

I just stood there looking stupid... watching the horses graze and Luisa cook under Jills optimistic gaze. I told Luisa what I'd seen; she didn't seem phased at all, and just had a bit of a chuckle saying something along the lines of... I didn't see it, but I certainly haven't ever heard a noise like that before; I didn't know it then, but we would hear it again within days though never again seeing it. We were soon a bit drunk, laughing by the campfire made with the little wood we had managed to carry with us from the last stand of low scrub some 5 or 6 km away, joking about aliens under that amazing star filled sky as our boys quietly grazed at the edge of the campfire light.

I was a bit on edge after the weird sighting, but I'd seen strange lights in the wilderness many times before, just never a "craft" as such. As I drifted off to sleep, I left my torch on my head and my hand rested on my bow; strung and ready to shoot. Lu was out to it like a light, but as always, I slept light and it wasn't long before Billy woke me. He wasn't startled, but he was aggressively letting the seemingly empty and silent steppe know he was a stallion, as he stamped his feet and huffed defiantly.

I slipped my boots on, leaving them undone and snuck silently from the tent in my long johns, slipping my puffer jacket on as I exited and notched an arrow. I had another few in my left hand as I tightly gripped my bow and switched on the bright LED headlamp to scan the steppe surrounding us. I knew there wouldn't be little green men or another silver disc, and what it most likely was terrifies me more than anything else on earth. I knew it would be the grey ghosts of the steppe, and I knew if they wanted to take us or the horses then there was very little I could do to prevent it.

Billy and Jac stood silently now, with both of them frozen like statues looking down into the bottom of the small depression we had camped in.

I swung the tiny beam of light through the darkness, feeling a chill run up my neck knowing unseen eyes were upon me now. I slowly walked in the direction the boys were staring, drawing slightly back on the heavy Mongolian horse bow armed with its single razor-sharp broad head-hunting arrow. I felt a lump in my throat; realizing how optimistic the idea of actually hitting a wolf with the bow was, and that in all likely hood I'd just make it angrier; but I knew the mere fact I was advancing, not retreating as I searched for the unseen predators was enough to hopefully drive them away.

I was right; catching a fleeting glimpse of three sets of eyes as they paused right on the edge of my torchlights limit, before vanishing into the inky blackness of the moonless night. I clicked my head torch off and smiled, though now shivering a little; it was freezing cold, and a night lit only by the brilliance of the unobscured star light bending over a black empty landscape. I smiled as I walked back to my sleeping partner, dog and horses who I could see were now happily grazing once more.

I smiled because an ex told me I would never be happy until I found a glorious death doing something stupid. Nah, she was way off. I smiled because I was living wild and free doing something that fulfilled my lust for life. I thought of all of the times I'd had nights like this; nights were I just knew there would be no way I could explain to a "normal person", what seeing and feeling things like this is like.

I dropped my bow by the tent, and wandered towards billy who softly murmured his need for affection in the darkness. I walked towards his black silhouette cut into a star filled backdrop and smiled as he lowered his head so I could stroke his face, scratch behind his ears and feel close to him. Despite his fur being cold to the touch, he nuzzled into me and I felt his warm breath against my face as he softly and kindly acknowledged my role in our partnership here on the steppe.

We were the same. Two gentle souls; capable of acts of great defiance; and if called for violence in the face of those that would seek to harm our family.

As Billy returned to grazing, I walked back towards bed, taking care not to trip over my shoe laces I'd still not done up, and again I started smiling. I may have even had a little bit of a chuckle softly to myself as I thought of all the people that had called me crazy over the years, saying I was going to die out here doing something like this.

Everyone dies. But how many of us really live?

They are the crazy ones; not me, not Luisa either.

Man is destined to failure if we live our whole lives being who others or society expects us to be. And for good or bad; Luisa and I have spent our lives just trying to be ourselves. I feel out here, in this empty freezing wilderness, we have found a way to feel whole. I was happy, and as Jill grumbled in disapproval as I moved her gently from my side of the bed, I looked forward to another day on the steppe with my family. I have no doubt when my time does come, my headstone will say nothing more than my regimental number, name, unit, and...

"I hoped for nothing, feared nothing, and I was free."

Your next step...

I hope that speaking so openly and honestly of our failures, the lessons learned and successes will inspire others to step from that world of "Kardashians" and cruelty at home, and into the uncertainty of an exciting day of horseback adventure in strange new lands.

One only needs to be a fly on the wall in any equestrian group to see the nastiness and petty bullying that's permeating every corner of Facebook. Sadly, it's usually the tall poppies that are bullied with the most vigorous and personal of assaults, and Luisa has on more than one occasion been in tears as a result of personal attacks. She's been hit with advice from armchair experts the world over, as have I; yet here we are, and they're at home. Everything from

"you shouldn't have your bridle over your head collar, the way people ride is cruel using whips, the horses are too thin."

Too worst of all....

"you have no place doing this, you're not experienced enough."

Well, hate all you want; she's living her best life and she's not hurting anyone. She wakes each day to the sound of horses grazing nearby, in the arms of a man that loves her with her little dog rudely trying to steal more of the sleeping bag. But who cares what they think?

Luisa and I both care how you feel. Are you happy? I hope so; I hope you're sitting at home on a sunny day in a cosy chair outside having not put this book down in a long while. I hope you're looking at an empty bottle of wine, a half-eaten wheel of brie, some sliced meat, olives sitting on a table beside you, having just brushed cracker crumbs off yourself as you get up for another glass of wine. You could be an awkward teen reading this in a treehouse, a veteran who like me didn't sleep last night, turning over things in his head there's no way he can ever change or un-live.

No matter who you are, Luisa and I hope you feel good about yourself having read our story, and hope you feel that there is a shot just waiting for you to take if you want it. It could be in any direction, in any field not just some crazy horse ride. There's always a chance to do something beautiful with your life; it doesn't always need a ton of money, perfect fitness, or amazing skills. It needs heart, and sometimes a little nudge from someone just like you, that took a leap they never dreamed they would.

We wrote this book not to tell the world how great we are, but to let others know how great they can be when they try. Much love to you all, Luisa, Jill and Pete.

"Life moves pretty fast, if you don't stop and look around once in a while, you could miss it."
Ferris Bueller

Acknowledgements.

There are a lot of people we would like to thank as well as recommend; and once more I would like to state that we are not paid, associated with in anyway, or interested in either financial sponsorship or material affiliation with a company. Objectivity and integrity is the goal for us in every recommendation we make, and if we are paid to promote products, we loose both of these.

Our Youtube videos were made by us, for us, and our family and friends, and will never be moniterised as a result. The songs used in our video's are the songs we listened to at the time; and we will continue to support and promote Aussie hip hop and the musicians that inspire us, not other people we're try to impress. We waited a while before we decided to make a Facebook page for people to follow our adventures on, mainly so we had the time to establish a little personal credibility.

The Youtube videos have been around since the start, and I must warn viewers many contain swearing, drinking and partying, excessive laughter and us being absolute dorks. We are real in every sense of the word and do not censor ourselves or our failures to look cool, hence our going joke... Be Professional. Look cool. When in all reality we struggle to do either. If you would like to follow our continuing adventures we can be found by searching the following.

YouTube: https://youtube.com/c/beprofessionallookcool
Instagram: https://Instagram.com/beprofessionallookcool
Facebook: https://www.facebook.com/beprofessionallookcool

Thank you to everyone who has been a part of this journey from its very conception to its random stops along the steppe and the amazing support we found in Georgia during the "COVID Pause". There are simply too many people to name individually but there is one person who needs very special thanks here and that is Alison Cheetham. Ali has been an incredible support to us all along the way but in particular in the final editing of this book. It's thanks to her alone. that this book is written in coherent english. A further edit was done by our mate Gary Greyling too and was much appreciated. Luisa and I are poor, and this help we couldn't afford otherwise has been huge for us both.

People and products we recommend for a long ride.

We have been meaning to do a video on our gear for a long time, and after over a year riding full time, we feel we have made enough mistakes to offer reasonably sound advice. Check out our Youtube channel for a full gear list and explanation. The recommendations we make below, we both stand by 100%, although there are no doubt many other great services and products out there we have yet to use or discover. We do not recommend specific equestrian clothing as we have found none of it is durable enough to survive a long ride where gear literally cannot fail you.

Ray Mears.
Ray Mears is the finest Outdoorsman of his kind in our opinion, and it is our distinct honor to have permission to recommend his youtube channel, bushcraft courses, and products.

Web: www.raymears.com
YouTube: htttps://www.youtube.com/user/RayMearsBushcraft
Facebook: https://www.facebook.com/raymearswoodlore/

MeatEater.
Stars In The Sky: A Hunting Story, is new to Netflix and is truly and emotional watch, although Steve Rinella's show "MeatEater" has been around for many seasons. Steve is articulate, professional and upholds the highest ethical standards in all his hunts and fishing trips. His web page is full of both great stories, recipes and links to podcasts and videos on outdoor pursuits.

Web: https://www.themeateater.com
Facebook: https://www.facebook.com/StevenRinellaMeatEater/

Woodpeckers. Mongolia.
If your looking to have an adventure like ours, and budget is something to seriously factor in, this is a great place to start. They run tours, as well as have affordable backpacker accommodation.

Email: woodpeckersinntours@gmail.com
Facebook: https://www.facebook.com/welovewoodpeckertours/

Dunedin riding centre. New Zealand.
It would be wrong not to mention Victoria Watt and her riding centre in the deep south of New Zealand. It was fixing the fences, cutting down trees and tidying paddocks there were I first started to get the idea to just run away on horseback.

Email: dunedinriding@yahoo.co.nz
Facebook: https://www.facebook.com/DunedinRidingCentre

Ghvevi horse riding. Georgia.
Tatia and her team are just the nicest, most authentic bunch of people able to offer you a real "Georgian" experience your able to see in many of our video's.

Email: ghvevihorseriding@gmail.com
Facebook: https://www.facebook.com/horseridingtbilisi/
Instagram: https://Instagram.com/horseridingtbilisi

Lost Ridge Ranch, and Living roots travel. Georgia.
Our time in Georgia was utterly amazing thanks in a really big way to John and his team. John changed the way I will view wine, as well as giving us alongside his amazing business partners and team some truly incredible experiences. Our trip would simply not have been what it was without these wonderful people, and should you find yourself in Georgia, they should be number one on your list.

Web: http://travellivingroots.com/horse-ranch/
Email: info@travellivingroots.com
Facebook: https://www.facebook.com/livingrootsranch/

Pheasants tears winery.
Pheasants tears wines are like nothing I've ever tasted, and I'm going to genuinely going to struggle without it in my life on a daily basis. If you can find it in a good wine stockist near you, treat yourself; you wont regret it!

Web: https://www.pheasantstears.com/

Breanna Wilson.
Breanna Wilson is a freelance writer, photographer, and adventurer who also happens to host some of the raddest tours in Mongolia. We met in Georgia, rode together, had many a wine together and got to know each other very well. She is the real deal; she can ride horses, motorbikes, is a straight up gangster, and has her finger on the pulse when it comes to adventure travel! She can line up motorbike tours of Mongolia among other adventures, and we would without recommend getting in touch if your planning a visit to Mongolia.

Web: https://meanwhileinmongolia.com/
Instagram: https://Instagram.com/breannajwilson

Tanja Mitton - Equestrian Success and Mindset.

There are two horsemen we need to mention here, the first is Tana Mitton. Aside from Pete, Tanja is the only person that has ever talked to Luisa, and trained her in regards to the mindset a rider requires to ride well. Luisa was very fortunate in having the opportunity to stay with her during her time in Australia, it wasn't the lessons themselves, but the way she taught Luisa to be aware of her thoughts that opened a door, and allowed her to develop in her own ways as a horseman. Luisas will be forever grateful for this amazing friend's influence in her life, and she runs clinics and lessons in Australia.

Email: tanja@tanjamitton.com
Facebook: https://www.facebook.com/tanjamitton69

Lorrie Duff – Liberty lane farms.

"My dad taught me to be a good person, and I just applied that to horses..." just one of the many things Pete recalls Lorrie telling him over way to much vodka and beer one night on the Mongolian steppe. Like Tanja did for Luisa, Lorrie didn't really give Pete lessons, but through conversation opened mental doors that allowed him to explore horsemanship in his own way.

Web: http://www.libertylanefarm.net/
Email: dufflorie@yahoo.com
Facebook: https://www.facebook.com/libertylanefarm/

Icebreaker merino clothing.

As mentioned at length, there is no finer base layer than merino. Icebreaker are both ethical in how they source their wool, and create long lasting garments of an exceptional standard and fit. Again, not exactly equestrian specific garments, but their value in warmth provided for compact size is second to none.

Web: https://www.icebreaker.com

Swazi clothing.
At the very top of any list of recommendations regarding outdoor gear, is one name that for us stands alone above all others; and that's Swazi. Swazi gear is worth every cent; and if your not sure about anything just call them, message them over facebook, instagram, or email. They stand buy their gear and so do we, as it's been tested and abused by people like us for years on end in some of the worlds harshest environments. It hasn't let us down, even after up to ten years of abuse, and even if it did, we could just send it back to have it fixed or replaced for free. We happily pay full retail price for their gear and proudly support Swazi and their commitment to quality.

Web: https://www.swazi.co.nz

Sitka clothing.
Like Swazi, Sitka clothing is built to last and stands up to the brutal punishment I give it daily like no other outdoors gear can. It's exceptionally comfortable, functional and although not cheap, it will outlast cheaper garments by such a degree its false economy to buy anything less than Sitka. Although like Swazi it is not equestran specific, it is the best gear for the environmental challenges you face on a long distance horse back adventure, rather than an afternoon in the arena on an english saddle.

Web: https://www.sitkagear.com/

Jet Boil stoves.
We swear by Jet Boil stoves. If you buy one, having never owned one... you'll ask yourself why it took you this long to switch after the first time you use it. Check them out at...

Web: https://www.jetboil.com

Gas bottle adapters.
But bare in mind in much of Asia the screw type gas canisters are hard to find but a adapter can easily be bought online, and Ive included both a link to an example of one, and also a video that will show you how to use it. In all honesty, genuine jet boil cans work best by far, but if you cant get them, this is the best solution.

YouTube: https://www.youtube.com/watch?v=k3fUJncqTaw
Web: https://www.amazon.com/Convert-Adapter-Nozzle-Cartridge-Canister/dp/B00U2B7UYQ

Speedy stitcher.
The speedy stitcher sewing awl is a simple to use item essential for the in field repair and maintenance of saddles, bags, boots, head collars and clothing. It can stitch through very thick leather and canvass and I would quite literally rate them as essential for a long ride and maintenance of kit in the field. Ivc included a video on how to use them, and you can purchase them in a good outdoors store, or on line at...

Web: http://www.speedystitcher.com
Youtube: https://www.youtube.com/watch?v=c6OatKzjeD0

Impact Gel saddle blankets and pads.
I am a firm believer the reason I have had no issue with both my back, and that of the horses I've ridden is due to the superb comfort provided by the two impact gel products I use. Neither were cheap; but if the saddle blanket is half as good for the horse, as the pad that sits under my bony butt is, then I know I am treating my horses with the most respect and kindness I possibly can. You can find them in all good horse shops or online at...

Web: https://impactgel.com/

Horseland Mornington.

As we've mentioned many times, there are many aspects of the horse world we don't particularly like; and when I told people of my plans I was often dismissed. But the staff of Horseland "a chain store in Australia" in Mornington treated me with both curiosity, and absolute professionalism at every dealing we had. I would recommend them without reservation, and their gear is of a very high and affordable standard, but it is the staffs attitude and customer focus that has their name in our acknowledgements.

Web: https://www.horseland.com.au
Facebook: https://www.facebook.com/hlandmornington

Emma Schmitt fine art. "Cover artist"

We met Emma over facebook on a horse related group around a year ago. At the time we didn't know she was an artist, but chatted over our mutual love of horses. It wasn't until we started to notice her online posts, that we came to realize that Emma, a classically trained artist who lives with her partner on a houseboat in the UK; was in fact an incredible artist. Her anthropomorphic portraits of animals combine both exceptional use of color, with superb realism, creating truly emotional images. Luisa and I both have tattoo's drawn by Emma, and are proud and humbled to share her work as the books cover image. She is available for commissioned works, and can be reached at...

Instagram: https://Instagram.com/emmaschmittart
Facebook: http://www.facebook.com/emmaschmittfineart

Quotes.
All Quotes were sourced from the individuals themselves or www.brainyquote.com.

Final Note...

Our new adventures are well underway as this goes to print, and believe us when we say... the best is yet to come!

Luisa, Pete and Jill.

Printed in Great Britain
by Amazon